P9-DBL-260

Electrical Power Systems Quality

Electrical Power Systems Quality

Roger C. Dugan
Senior Consultant, Electrotek Concepts, Inc.

Mark F. McGranaghan
General Manager, Electrotek Concepts, Inc.

H. Wayne Beaty
Managing Editor, Electric Light & Power

McGraw-Hill

New York San Francisco Washington, D.C. Auckland Bogotá
Caracas Lisbon London Madrid Mexico City Milan
Montreal New Delhi San Juan Singapore
Sydney Tokyo Toronto

Library of Congress Cataloging-in-Publication Data

Dugan, Roger C.
Electrical power systems quality / Roger C. Dugan, Mark F. McGranaghan, H. Wayne Beaty.
p. cm.
Includes bibliographical references and index.
ISBN 0-07-018031-8
1. Electric power system stability. 2. Electric power systems—Quality control. 3. Electric power-plants—Quality Control.
I. McGranaghan, M. F. II. Beaty, H. Wayne. III. Title.
TK1010.D84 1996
621.319'1—dc20 95-45698
CIP

McGraw-Hill
A Division of The ***McGraw-Hill*** *Companies*

2 3 4 5 6 7 8 9 0 DOC/DOC 9 0 0 9 8 7

ISBN 0-07-018031-8

The sponsoring editor for this book was Harold B. Crawford, the editing supervisor was Stephen M. Smith, and the production supervisor was Donald F. Schmidt. It was set in Century Schoolbook by Ron Painter of McGraw-Hill's Professional Book Group composition unit.

Printed and bound by R. R. Donnelley & Sons Company.

This book is printed on recycled, acid-free paper containing a minimum of 50% recycled de-inked fiber.

Contents

Foreword

Marek Samotyj*

The Importance of Power Quality

I want to commend the authors on this valuable contribution to an area that continues to grow in importance. At the Electric Power Research Institute, we have been studying power quality problems and solutions for over ten years. Electrotek Concepts has been the contractor for much of this research, including over 50 power quality case studies at end-user facilities. Much of the knowledge developed from these case studies and other power quality research on both sides of the meter has been included in this book. The information should provide power quality engineers with a great head start in evaluating power quality concerns from either the utility perspective or the end-user perspective.

Power quality includes such a broad range of concerns and evaluations that it is very difficult to organize the material in any cohesive manner. The evaluations can include everything from transmission system fault studies to transient voltage surge suppression for computer data lines. To make the problem even more difficult, these concerns need to be explained so that they can be understood by electric utility engineers, end-user facility managers, and equipment designers. No easy job.

This book should help contribute to the overall understanding of power quality problems and how they involve interaction over

*Marek Samotyj is Manager of the Power Quality Business Unit at the Electric Power Research Institute in Palo Alto, California. He has been responsible for many of the power quality case studies and technology transfer activities that have influenced the industry in the last ten years. He has sponsored international power quality conferences since 1988 and has published numerous papers, technical articles, and reports dealing with all aspects of power quality. The power quality case studies sponsored by EPRI and participating electric utilities have resulted in a tremendous knowledge base and development of new solutions to power quality problems. This information is currently being incorporated into the first interactive power quality database in the industry.

such a wide range of the system. This can only help in the process of getting everyone to work together in developing solutions to the problems. Equipment designers must work to make equipment that is compatible with the real-world power system. This might mean lower levels of harmonic generation or better voltage sag ride-through characteristics. Facilities managers and designers must build and operate systems that take into account the interaction issues between the end-user facilities and the power system. For instance, power-factor correction should be coordinated with harmonic control requirements to avoid system resonance. Finally, electric utility engineers must understand the sensitivity and characteristics of the end-use equipment. These characteristics may influence the system design, from protection practices to capacitor switching procedures.

I think we are making tremendous progress with this whole concept of "compatibility." I am Secretary of the IEEE Power Quality Standards Coordinating Committee, SCC-22. This committee coordinates standards development for a wide variety of organizations interested in the concept of compatibility. We are even starting to get coordination between IEEE and international standards organizations, such as IEC. I continue to look forward to the day when all of the power quality standards are international in scope and include input from all interested parties. This book will help establish an understanding of the important issues, which should help make this happen.

Acknowledgments

This book is a collection of material on electrical systems power quality compiled by Roger C. Dugan and Mark F. McGranaghan of Electrotek Concepts, Inc., from contributions of Electrotek's staff. H. Wayne Beaty served as editor for the book. We are indebted to many of the staff of the Power Systems Engineering group of Electrotek for their various contributions. It is not practical in all cases to identify each person's specific contribution because of how we merged the material into cohesive text. Therefore, we would like to generally acknowledge the efforts of the people whose names are listed below along with their areas of expertise relevant to this book.

Daniel Brooks is Power Systems Engineer. He is an investigator on the Electric Power Research Institute (EPRI) Distribution Power Quality Assessment project (RP 3098-1) and served as a proofreader for this book.

Rory V. Dwyer is Senior Power Systems Engineer. He works with utility and industrial engineers to protect equipment from harmonics, transients, and voltage sags.

Thomas E. Grebe is Manager of Utility Studies. He is Project Manager for the EPRI Distribution Power Quality Assessment project and manages the EMTP and HARMFLO Users Groups.

Erich W. Gunther is Director of Technology. He developed the SuperHarm® and TOP® computer programs for harmonics and transients analysis and is liaison with CIGRE and CIRED power quality activities for both SCC-22 and P1159.

Afroz Khan is Power Systems Engineer. She has been involved in the analysis of the effects of the harmonics produced by compact fluorescent lamps on utility distribution systems and the support of the SuperHarm® computer program.

Jack A. King is Manager of Software Development. He has developed numerous computer programs for power quality analysis, including the PASS® software for the BMI 8010 PQNode®.

Jeffrey D. Lamoree is Manager of Power Quality Projects. He directs power quality monitoring and analysis projects involving interaction between utility and end-user systems. He is the manager of a project of EPRI to develop a power quality database that will incorporate the results from numerous case studies, monitoring efforts, and equipment testing.

Christopher J. Melhorn is Supervisor of Industrial Applications. He coordinates projects involving power quality monitoring and simulations for industrial and commercial customers. He has developed software for monitoring instruments, including the FlukeView® software for the Fluke 97 and the Fluke 41.

David R. Mueller is Senior Consultant. He has performed large power quality studies with particular focus on industrial power quality concerns such as voltage sags. He has spent two years working with East Midlands Electricity (England) to develop a power quality services (PQS) business unit and help the utility start an extensive power quality monitoring project.

D. Daniel Sabin is Senior Power Systems Engineer. He is currently an investigator on the EPRI Distribution Power Quality Assessment project, responsible for analyzing the power quality data being collected at nearly 300 monitoring points on utility distribution feeders throughout the United States.

J. Charles Smith is Vice President and General Manager of International and Renewable Programs. He has been instrumental in the establishment of power quality standards and served as chairman of the P1159 Working Group.

Robert M. Zavadil is Manager of Utility Applications. He is an expert in adjustable-speed drive application and is also active in renewable energy sources.

A significant portion of the background material for this book was developed through numerous power quality case studies. Most of these were sponsored by EPRI and individual participating utilities. The authors would like to particularly acknowledge the contribution of Marek Samotyj of EPRI. Marek was the EPRI Project Manager for all of these investigations and has been a major force in advancing power quality research. In particular, the international power quality conferences sponsored by EPRI from 1990 to 1995 have provided valuable information to the entire industry.

We would also like to express our appreciation to the equipment suppliers who so graciously supplied us with photographs of their products for use in this book: Cooper Power Systems, ABB Power T&D Co., Square D Co., Joslyn High-Voltage Corporation, Fluke Corporation, and BMI.

Electrical Power Systems Quality Home Page

A home page on the World-Wide Web (WWW) has been established for readers of this book. Readers are invited to submit comments and participate in discussions related to the book by connecting their WWW browsers to

http://www.electrotek.com/pqbook/

or sending e-mail to

pqbook@electrotek.com

Roger C. Dugan
Mark F. McGranaghan
H. Wayne Beaty

Chapter

1

Introduction

Both electric utilities and end users of electrical power are becoming increasingly concerned about the quality of electric power. The term *power quality* has become one of the most prolific buzzwords in the power industry since the late 1980s. It is an umbrella concept for a multitude of individual types of power system disturbances. The issues that fall under this umbrella are not necessarily new. What is new is that engineers are now attempting to deal with these issues with a systems approach rather than as individual problems.

There are four major reasons for the growing concern:

1. Load equipment is more sensitive to power quality variations than equipment applied in the past. Many new load devices contain microprocessor-based controls and power electronic devices that are sensitive to many types of disturbances.
2. The increasing emphasis on overall power system efficiency has resulted in a continued growth in the application of devices such as high-efficiency, adjustable-speed motor drives and shunt capacitors for power factor correction to reduce losses. This is resulting in increasing harmonic levels on power systems and has many people concerned about the future impact on system capabilities.
3. Increased awareness of power quality issues by the end users. Utility customers are becoming better informed about such issues as interruptions, sags, and switching transients and are challenging the utilities to improve the quality of power delivered.

4. Many things are now interconnected in a network. Integrated processes mean that the failure of any component has much more important consequences.

The main impetus behind these reasons is increased productivity for utility customers. Manufacturers want faster, more productive, more efficient machinery. Utilities encourage this effort because it helps their customers become more profitable and also helps defer large investments in substations and generation by using more efficient load equipment. Interestingly, the equipment installed to increase the productivity is also often the equipment that suffers the most from common power disruptions. And the equipment is sometimes the source of additional power quality problems.

1.1 What Is Power Quality?

There can be completely different definitions for power quality, depending on one's frame of reference. For example, utilities may define power quality as reliability and show statistics demonstrating that the system is 99.98 percent reliable. The manufacturer of load equipment may define quality power as those characteristics of the power supply that enable the equipment to work properly. These can be very different for different equipment and different manufacturers. However, power quality is ultimately a customer-driven issue and the customer's point of reference takes precedence. Therefore, the following definition of a power quality problem is used in this book:

> Any power problem manifested in voltage, current, or frequency deviations that results in failure or misoperation of customer equipment

There are many misunderstandings regarding the causes of power quality problems. The charts in Fig. 1.1 show the results of one survey conducted by the Georgia Power Company in which both utility and customer personnel were polled about what causes power quality problems. While surveys of other market sectors might indicate different splits between the categories, these charts clearly illustrate one common theme that arises repeatedly in such surveys: the utility and customer perspectives are often much different. While both tended to blame about two-thirds of the events on natural phenomena (e.g.,

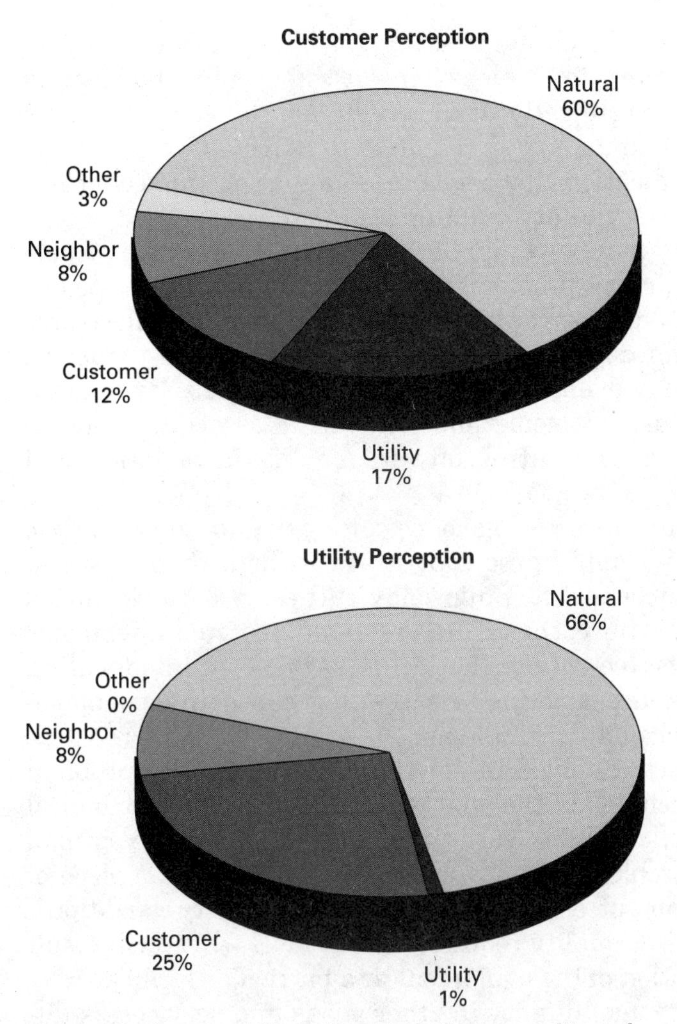

Figure 1.1 Survey results on the causes of power quality problems. (*Courtesy of Georgia Power Co.*)

lightning), the customers think that the utility is at fault much more frequently than utility personnel.

When there is a power problem with a piece of equipment, end users may be quick to complain to the utility of an "outage" or "glitch" that has caused the problem. However, the utility records may indicate no abnormal events on the feed to the customer. It must be realized that there are many events resulting in customer problems that never show up in the utility statistics.

One example is capacitor switching, which is quite common and normal on the utility system, but can cause transient overvoltages that disrupt manufacturing machinery. Another example is a momentary fault elsewhere in the system that causes the voltage to sag briefly at the customer in question. This might cause an adjustable-speed drive to trip off, but the utility will have no indication that anything is amiss on the feeder unless it has a power quality monitor installed.

In addition to real power quality problems, there are also perceived power quality problems that may actually be related to hardware, software, or control system malfunctions. Electronic components can degrade over time due to repeated transient voltages and may eventually fail due to a relatively low-magnitude event. Thus, it is sometimes difficult to associate a failure with a specific cause. Control software may not have anticipated a particular occurrence.

In response to this growing concern for power quality, electric utilities are developing programs that can help them respond to customer concerns. The philosophy of these programs ranges from *reactive,* where the utility responds to customer complaints, to *proactive,* where the utility is involved in educating the customer and promoting services that can help develop solutions to power quality problems.

The economics involved in solving a power quality problem must also be included in the analysis. It is not always economical to eliminate power quality variations. In many cases, the optimal solution to a problem may involve making a particular piece of sensitive equipment less sensitive to power quality variations. The level of power quality required is that level which will result in proper operation of the equipment at a particular facility.

Power quality, like quality in other goods and services, is difficult to quantify. There is no single accepted definition of "quality power." There are standards for voltage and other technical criteria that may be measured, but the ultimate measure of power quality is determined by the performance and productivity of end-user equipment. If the electric power is inadequate for those needs, then the "quality" is lacking.

Perhaps nothing has been more symbolic of a mismatch in the power delivery system and consumer technology than the "blinking clock" phenomenon. Clock designers created the blinking display of a digital clock to warn of possible incorrect time after loss of power and inadvertently created one of the first power quality monitors. It has made the homeowner aware that there are numerous minor disturbances occurring throughout the power delivery system, that may have no ill effects other than that they are detected by a clock. With many appliances now coming with a clock built in, the average household may have about a dozen clocks that must be reset when there is a brief interruption of power. Older technology motor-driven clocks would simply lose a few seconds during minor disturbances and then promptly come back into synchronism.

1.2 Power Quality = Voltage Quality

While the common term for describing the subject of this book is *power* quality, it is actually the quality of the voltage that is being addressed in most cases. Technically, in engineering terms, *power* is the rate of delivery of energy and is proportional to the product of the voltage and current. It would be difficult to define the quality of this quantity in any meaningful manner. The power supply system can only control the quality of the voltage; it has no control over the currents that particular loads might draw. Therefore, the standards in the power quality area are devoted to maintaining the supply voltage within certain limits.

Alternating current power systems are designed to operate at a sinusoidal voltage of a given frequency (typically 50 or 60 Hz) and magnitude. Any significant deviation in the magnitude, frequency, or purity of waveform is a potential power quality problem.

Of course, there is always a close relationship between voltage and current in any practical power system. Although the generators may provide a near-perfect sine-wave voltage, the current passing through the impedance of the system can cause a variety of disturbances to the voltage. For example,

1. The current resulting from a short circuit causes the voltage to sag, or disappear completely, as the case may be.
2. Currents from lightning strokes passing through the power system cause high impulse voltages that frequently flash over insulation and lead to other phenomena, such as short circuits.
3. Distorted currents from harmonic-producing loads also distort the voltage as they pass through the system impedance. Thus a distorted voltage is presented to other end users.

Therefore, while it is the voltage with which we are ultimately concerned, we must address phenomena in the current to understand the basis of many power quality problems.

1.3 Why Are We Concerned about Power Quality?

The ultimate reason that we are interested in power quality is economic value. There are economic impacts on utilities, their customers, and suppliers of load equipment.

The quality of power can have a direct economic impact on many industrial consumers. There has recently been a great emphasis on revitalizing industry with more automation and more modern equipment. This usually means electronically controlled, energy-efficient equipment which is often much more sensitive to deviations in the supply voltage than its electromechanical predecessors. Thus, as with the blinking clock in residences, industrial customers are now more acutely aware of minor disturbances in the power system. Big money is associated with these disturbances. It is not uncommon for a single, commonplace, momentary utility breaker operation to result in a $10,000 loss to an average-sized industrial concern by shutting down a production line that requires 4 h to restart.

The electric utility is concerned about power quality issues as well. Meeting customer expectations and maintaining customer confidence is a strong motivator. With today's movement toward competition between utilities, it is more important than ever. The loss of a disgruntled customer to a competing power supplier can have a very significant impact financially on a utility.

Besides the obvious financial impacts on both utilities and industrial customers, numerous indirect and intangible costs are associated with power quality problems. Residential customers typically do not suffer direct financial loss or the inability to earn income as a result of most power quality problems, but they can be a potent force when they perceive that the utility is providing poor service. The sheer number of complaints require utilities to provide staffing to handle them.

Also, public interest groups frequently intervene with public service commissions, requiring the utilities to expend financial resources on lawyers, consultants, studies, and the like to counter the intervention. While all of this is certainly not the result of power quality problems, a reputation for providing poor quality service does not help matters.

Load equipment suppliers generally find themselves in a very competitive market with most customers buying on lowest cost. Thus, there is a general disincentive to add features to the equipment to withstand common disturbances unless the customer specifies otherwise. Many manufacturers are also unaware of the types of disturbances that can occur on power systems. There is a need for education on this subject, which is one of the key purposes of this book.

The primary responsibility for correcting inadequacies in load equipment ultimately lies with the end user who must purchase and operate it. Specifications must include power performance criteria. Since many end users are also unaware of the pitfalls, one useful service that utilities can provide is dissemination of information on power quality and the requirements of load equipment to properly operate in the real world.

1.4 Who Should Use This Book

Power quality issues frequently cross the energy meter boundary between the utility and the end-user. Therefore, this book addresses issues of interest to both utility engineers and industrial engineers and technicians. Every attempt has been made to provide a balanced approach to the presentation of the problems and solutions. The book should also be of interest to designers of load equipment to learn about the environment in which their equipment must operate and the peculiar difficulties their customers might have trying to operate their equipment. It is to be hoped that this book will serve as common ground on which these three entities—utility, customer, and supplier—can meet to resolve problems.

This book is intended to serve both as a reference book and a textbook for utility distribution engineers and key technical personnel with industrial end users. Parts of the book are tutorial in nature for the newcomer to power quality and power systems, while other parts are strictly reference for the experienced practitioner.

1.5 Overview of the Contents

The chapters of the book are organized as follows:

Chapter 2 provides background material on the different types of power quality phenomena and describes some current activities to develop standard terms and definitions for power quality phenomena.

Chapters 3 through 6 are the heart of the book, describing four major classes of power quality variations in detail: sags and interruptions, transients, harmonics, and long-duration voltage variations.

Although, perhaps, not strictly a power quality subject, many power quality variations arise from wiring and grounding problems. Chapter 7 provides a concise summary of key wiring and grounding problems and gives some general guidance on identifying and correcting them.

Finally, Chap. 8 provides a guide for site surveys and power quality monitoring.

Chapter

2

Terms and Definitions

2.1 Need for a Consistent Vocabulary

The term *power quality* is applied to a wide variety of electromagnetic phenomena on the power system. The increasing application of electronic equipment has heightened the interest in power quality in recent years and this has been accompanied by the development of a special terminology to describe the phenomena. Unfortunately, this terminology has not been consistent across different segments of the industry. This has caused a considerable amount of confusion as vendors and end users have attempted to understand why the equipment is not working as expected.

Many ambiguous words have been used that have multiple or unclear meanings. For example, the term *surge* is used to describe a wide variety of disturbances that cause equipment failures or misoperation. A *surge suppressor* can suppress some of these, but will have absolutely no effect on others. Terms like *glitch* and *blink* that have no technical meaning at all have crept into the vocabulary.

This chapter describes a consistent terminology that can be used to describe power quality variations. We also explain why some commonly used terminology is inappropriate in power quality discussions.

2.2 General Classes of Power Quality Problems

The terminology presented here reflects recent U.S. and international efforts to standardize definitions of power quality terms. The Institute of Electrical and Electronics Engineers Standards Coordinating Committee 22 (IEEE SCC22) has led the main effort in the United States to coordinate power quality standards. It has the responsibilities across several societies of the IEEE, principally the Industry Applications Society and the Power Engineering Society. It coordinates with international efforts through liaisons with the International Electrotechnical Commission (IEC) and the Congress Internationale des Grand Réseaux Électriques a Haute Tension (CIGRE; in English, International Conference on Large High-Voltage Electric Systems).

The IEC classifies electromagnetic phenomena into the groups shown in Table 2.1.[1] We will be primarily concerned with the first four classes in this book.

The U.S. power industry efforts to develop recommended practices for monitoring electric power quality have added a few terms to the IEC terminology.[2] *Sag* is used as a synonym to the IEC term *dip*. The category *short-duration variations* is used to

TABLE 2.1 Principal Phenomena Causing Electromagnetic Disturbances as Classified by the IEC

Conducted low-frequency phenomena
Harmonics, interharmonics
Signal systems (power line carrier)
Voltage fluctuations
Voltage dips and interruptions
Voltage unbalance
Power frequency variations
Induced low-frequency voltages
dc in ac networks
Radiated low-frequency phenomena
Magnetic fields
Electric fields
Conducted high-frequency phenomena
Induced continuous-wave (CW) voltages or currents
Unidirectional transients
Oscillatory transients
Radiated high-frequency phenomena
Magnetic fields
Electric fields
Electromagnetic fields
Continuous waves
Transients
Electrostatic discharge phenomena (ESD)
Nuclear electromagnetic pulse (NEMP)

refer to voltage dips and short interruptions. The term *swell* is introduced as an inverse to sag (dip). The category *long-duration variation* has been added to deal with ANSI (American National Standards Institute) C84.1 limits. The category *noise* has been added to deal with broadband conducted phenomena.

The category *waveform distortion* is used as a container category for the IEC *harmonics, interharmonics,* and *dc in ac networks* phenomena as well as an additional phenomenon from IEEE standard 519 called *notching*.

Table 2.2 shows the categorization of electromagnetic phenomena used for the power quality community. The phenomena listed in the table can be described further by listing appropriate attributes. For steady-state phenomena, the following attributes can be used[1]:

- Amplitude
- Frequency
- Spectrum
- Modulation
- Source impedance
- Notch depth
- Notch area

For non-steady-state phenomena, other attributes may be required[1]:

- Rate of rise
- Amplitude
- Duration
- Spectrum
- Frequency
- Rate of occurrence
- Energy potential
- Source impedance

The table provides information regarding typical spectral content, duration, and magnitude where appropriate for each category of electromagnetic phenomena.[1,4,5] The categories shown in Table 2.2, when used with the attributes mentioned above, pro-

TABLE 2.2 Categories and Characteristics of Power System Electromagnetic Phenomena

Categories	Typical spectral content	Typical duration	Typical voltage magnitude
1.0 Transients			
1.1 Impulsive			
1.1.1 Nanosecond	5-ns rise	<50 ns	
1.1.2 Microsecond	1-μs rise	50 ns–1 ms	
1.1.3 Millisecond	0.1-ms rise	> 1 ms	
1.2 Oscillatory			
1.2.1 Low frequency	<5 kHz	0.3–50 ms	0–4 pu
1.2.2 Medium frequency	5–500 kHz	20 μs	0–8 pu
1.2.3 High frequency	0.5–5 MHz	5 μs	0–4 pu
2.0 Short-duration variations			
2.1 Instantaneous			
2.1.1 Interruption		0.5–30 cycles	<0.1 pu
2.1.2 Sag (dip)		0.5–30 cycles	0.1–0.9 pu
2.1.3 Swell		0.5–30 cycles	1.1–1.8 pu
2.2 Momentary			
2.2.1 Interruption		30 cycles–3 s	<0.1 pu
2.2.2 Sag (dip)		30 cycles–3 s	0.1–0.9 pu
2.2.3 Swell		30 cycles–3 s	1.1–1.4 pu
2.3 Temporary			
2.3.1 Interruption		3 s–1 min	<0.1 pu
2.3.2 Sag (dip)		3 s–1 min	0.1–0.9 pu
2.3.3 Swell		3 s–1 min	1.1–1.2 pu
3.0 Long-duration variations			
3.1 Interruption, sustained		>1 min	0.0 pu
3.2 Undervoltages		>1 min	0.8–0.9 pu
3.3 Overvoltages		>1 min	1.1–1.2 pu
4.0 Voltage unbalance		Steady state	0.5–2%
5.0 Waveform distortion			
5.1 dc offset		Steady state	0–0.1%
5.2 Harmonics	0–100th harmonic	Steady state	0–20%
5.3 Interharmonics	0–6 kHz	Steady state	0–2%
5.4 Notching		Steady state	
5.5 Noise	Broadband	Steady state	0–1%
6.0 Voltage fluctuations	<25 Hz	Intermittent	0.1–7%
7.0 Power frequency variations		<10 s	

vide a means to clearly describe an electromagnetic disturbance. The categories and their descriptions are important in order to be able to classify measurement results and to describe electromagnetic phenomena which can cause power quality problems.

2.3 Transients

The term *transients* has long been used in the analysis of power system variations to denote an event that is undesirable but momentary in nature. The notion of a damped oscillatory transient due to a RLC [resistance, inductance ("L" is the symbol for inductance), and capacitance] network is probably what most power engineers think of when they hear the word *transient.*

Other definitions in common use are broad in scope and simply state that a transient is "that part of the change in a variable that disappears during transition from one steady state operating condition to another."[8] Unfortunately, this definition could be used to describe just about anything unusual that happens on the power system.

Another word in common usage that is often considered synonymous with transient is *surge.* A utility engineer may think of a surge as the transient resulting from a lightning stroke for which a surge arrester is used for protection. End users frequently use the word indiscriminately to describe anything unusual that might be observed on the power supply ranging from sags to swells to interruptions. Because there are many potential ambiguities with this word in the power quality field, we will generally avoid using it unless we have specifically defined what it refers to.

Broadly speaking, transients can be classified into two categories, *impulsive* and *oscillatory.* These terms reflect the waveshape of a current or voltage transient. We will describe these two categories in more detail.

2.3.1 Impulsive transient

An *impulsive transient* is a sudden, non–power frequency change in the steady-state condition of voltage, current, or both, that is unidirectional in polarity (primarily either positive or negative).

Impulsive transients are normally characterized by their rise and decay times, which can also be revealed by their spectral content. For example, a 1.2×50-μs 2000-V impulsive transient

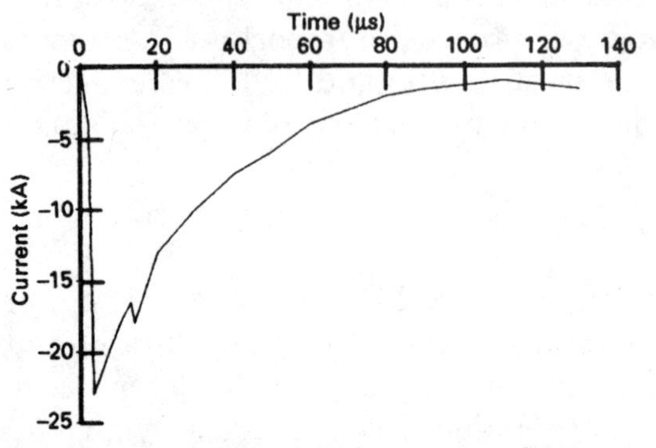

Figure 2.1 Lightning stroke current impulsive transient.

nominally rises from zero to its peak value of 2000 V in 1.2 μs, then decays to half its peak value in 50 μs.

The most common cause of impulsive transients is lightning. Figure 2.1 illustrates a typical current impulsive transient caused by lightning.

Due to the high frequencies involved, the shape of impulsive transients can be changed quickly by circuit components and may have significantly different characteristics when viewed from different parts of the power system. They are generally not conducted far from the source where they enter the power system, although they may, in some cases, be conducted for quite some distance along utility lines. Impulsive transients can excite the natural frequency of power system circuits and produce oscillatory transients.

2.3.2 Oscillatory transient

An *oscillatory transient* is a sudden, non–power frequency change in the steady-state condition of voltage, current, or both, that includes both positive and negative polarity values.

An oscillatory transient consists of a voltage or current whose instantaneous value changes polarity rapidly. It is described by its spectral content (predominate frequency), duration, and magnitude. The spectral content subclasses defined in Table 2.1 are high, medium, and low frequency. The frequency ranges for these classifications are chosen to coincide with common types of power system oscillatory transient phenomena.

Oscillatory transients with a primary frequency component greater than 500 kHz and a typical duration measured in mi-

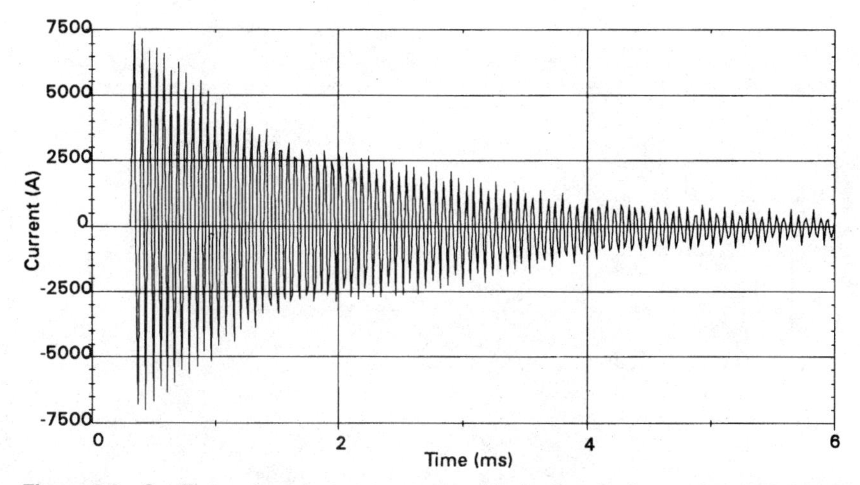

Figure 2.2 Oscillatory transient current caused by back-to-back capacitor switching.

croseconds (or several cycles of the principal frequency) are considered high-frequency oscillatory transients. These transients are often the result of a local system response to an impulsive transient.

A transient with a primary frequency component between 5 and 500 kHz with duration measured in the tens of microseconds (or several cycles of the principal frequency) is termed a *medium-frequency transient.*

Back-to-back capacitor energization results in oscillatory transient currents in the tens of kilohertz as illustrated in Fig. 2.2. Cable switching results in oscillatory voltage transients in the same frequency range. Medium-frequency transients can also be the result of a system response to an impulsive transient.

A transient with a primary frequency component less than 5 kHz, and a duration from 0.3 to 50 ms, is considered a *low-frequency transient.*

This category of phenomena is frequently encountered on utility subtransmission and distribution systems and is caused by many types of events. The most frequent is capacitor bank energization, which typically results in an oscillatory voltage transient with a primary frequency between 300 and 900 Hz. The peak magnitude can approach 2.0 pu (per unit), but is typically 1.3 to 1.5 pu with a duration of between 0.5 and 3 cycles depending on the system damping (Fig. 2.3).

Oscillatory transients with principal frequencies less than 300 Hz can also be found on the distribution system. These are

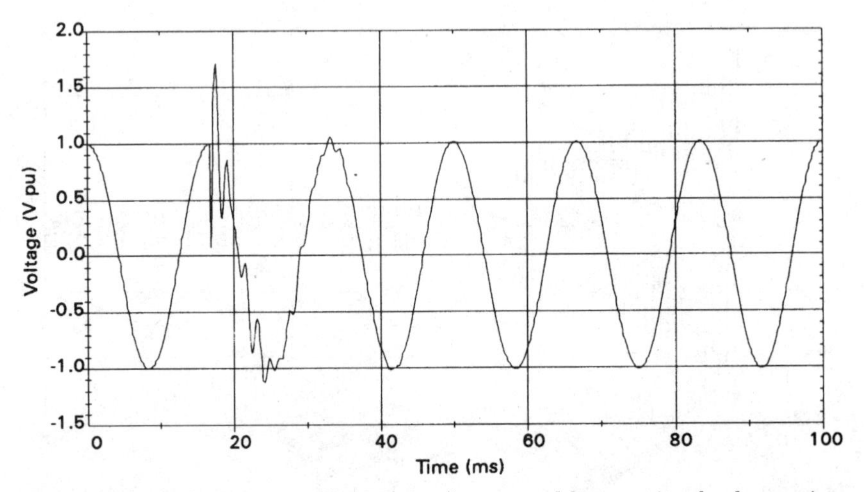

Figure 2.3 Low-frequency oscillatory transient caused by capacitor bank energization.

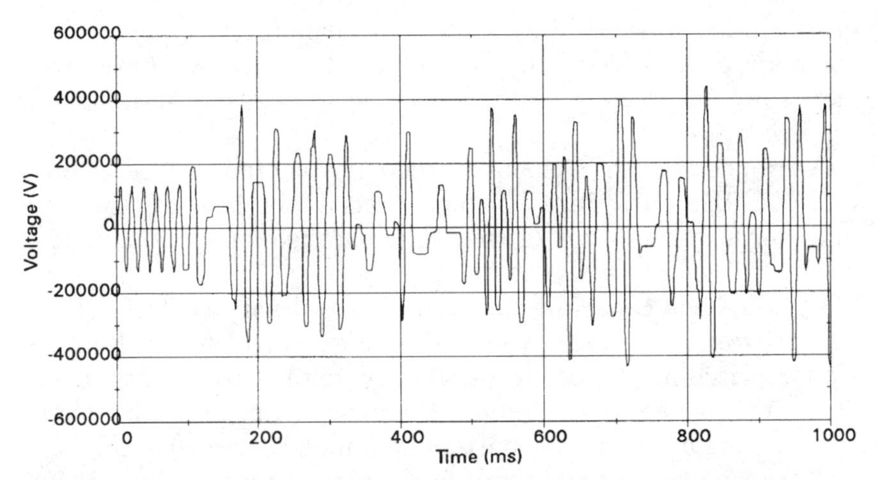

Figure 2.4 Low-frequency oscillatory transient caused by ferroresonance of an unloaded transformer.

generally associated with ferroresonance and transformer energization (Fig. 2.4). Transients involving series capacitors could also fall into this category. They occur when the system responds by resonating with low-frequency components in the transformer inrush current (second and third harmonic) or when unusual conditions result in ferroresonance.

It is also possible to categorize transients (and other disturbances) according to their *mode*. Basically, a transient in a three-phase system with a separate neutral conductor can be either *common mode* or *normal mode,* depending on whether it appears between line or neutral and ground, or between line and neutral.

2.4 Long-Duration Voltage Variations

Long-duration variations encompass root-mean-square (rms) deviations at power frequencies for longer than 1 min. ANSI C84.1 specifies the steady-state voltage tolerances expected on a power system. A voltage variation is considered to be long duration when the ANSI limits are exceeded for greater than 1 min.

Long-duration variations can be either *overvoltages* or *undervoltages.* Overvoltages and undervoltages generally are not the result of system faults, but are caused by load variations on the system and system switching operations. Such variations are typically displayed as plots of rms voltage versus time.

2.4.1 Overvoltage

An *overvoltage* is an increase in the rms ac voltage greater than 110 percent at the power frequency for a duration longer than 1 min.

Overvoltages are usually the result of load switching (e.g., switching off a large load, or energizing a capacitor bank). The overvoltages result because the system is either too weak for the desired voltage regulation or voltage controls are inadequate. Incorrect tap settings on transformers can also result in system overvoltages.

2.4.2 Undervoltage

An *undervoltage* is a decrease in the rms ac voltage to less than 90 percent at the power frequency for a duration longer than 1 min.

Undervoltages are the result of the events which are the reverse of the events that cause overvoltages. A load switching on or a capacitor bank switching off can cause an undervoltage until voltage regulation equipment on the system can bring the voltage back to within tolerances. Overloaded circuits can also result in undervoltages.

The term *brownout* is often used to describe sustained periods of undervoltage initiated as a specific utility dispatch strategy to reduce power demand. Because there is no formal definition for brownout, and it is not as clear as the term undervoltage when trying to characterize a disturbance, the term *brown- out* should be avoided.

2.4.3 Sustained interruptions

When the supply voltage has been zero for a period of time in excess of 1 min, the long-duration voltage variation is considered a *sustained interruption.* Voltage interruptions longer than 1 min are often permanent and require human intervention to repair the system for restoration. The term *sustained interruption* refers to specific power system phenomena and, in general, has no relation to the usage of the term *outage.* Utilities use *outage* or *interruption* to describe phenomena of similar nature for reliability reporting purposes. However, this causes confusion for end users who think of an outage as any interruption of power that shuts down a process. This could be as little as one-half of a cycle.

The term *outage,* as defined in IEEE Standard 1008, does not refer to a specific phenomenon, but rather to the state of a component in a system that has failed to function as expected. Also, use of the term *interruption* in the context of power quality monitoring has no relation to reliability or other continuity of service statistics. Thus, this term has been defined to be more specific regarding the absence of voltage for long periods.

2.5 Short-Duration Voltage Variations

This category encompasses the IEC category of Voltage Dips and Short Interruptions. Each type of variation can be designated as *instantaneous, momentary,* or *temporary,* depending on its duration as defined in Table 2.2.

Short-duration voltage variations are caused by fault conditions, the energization of large loads which require high starting currents, or intermittent loose connections in power wiring. Depending on the fault location and the system conditions, the fault can cause either temporary voltage drops (*sags*), or voltage rises (*swells*), or a complete loss of voltage (*interruptions*). The fault condition can be close to or remote from the point of interest. In either case, the impact on the voltage during the actual fault condition is of short-duration variation until protective devices operate to clear the fault.

2.5.1 Interruption

An *interruption* occurs when the supply voltage or load current decreases to less than 0.1 pu for a period of time not exceeding 1 min.

Interruptions can be the result of power system faults, equipment failures, and control malfunctions. The interruptions are measured by their duration since the voltage magnitude is always less than 10 percent of nominal. The duration of an interruption due to a fault on the utility system is determined by the operating time of utility protective devices. Instantaneous reclosing generally will limit the interruption caused by a nonpermanent fault to less than 30 cycles. Delayed reclosing of the protective device may cause a momentary or temporary interruption. The duration of an interruption due to equipment malfunctions or loose connections can be irregular.

Some interruptions may be preceded by a voltage sag when these interruptions are due to faults on the source system. The voltage sag occurs between the time a fault initiates and the protective device operates. Figure 2.5 shows such a momentary interruption during which voltage sags to about 20 percent for about three cycles and then drops to zero for about 1.8 s until the recloser closes back in. The sag waveform is typical of an arcing fault.

2.5.2 Sags (dips)

A *sag* is a decrease to between 0.1 and 0.9 pu in rms voltage or current at the power frequency for durations from 0.5 cycles to 1 min.

The power quality community has used the term *sag* for many years to describe a short-duration voltage decrease. Although the term has not been formally defined, it has been increasingly accepted and used by utilities, manufacturers, and end users. The IEC definition for this phenomenon is *dip*. The two terms are considered interchangeable, with *sag* being the preferred synonym in the U.S. power quality community.

Terminology used to describe the magnitude of a voltage sag is often confusing. A "20 percent sag" can refer to a sag which results in a voltage of 0.8 pu, or 0.2 pu. The preferred terminology would be one that leaves no doubt as to the resulting voltage level: "a sag to 0.8 pu" or "a sag whose magnitude was 20

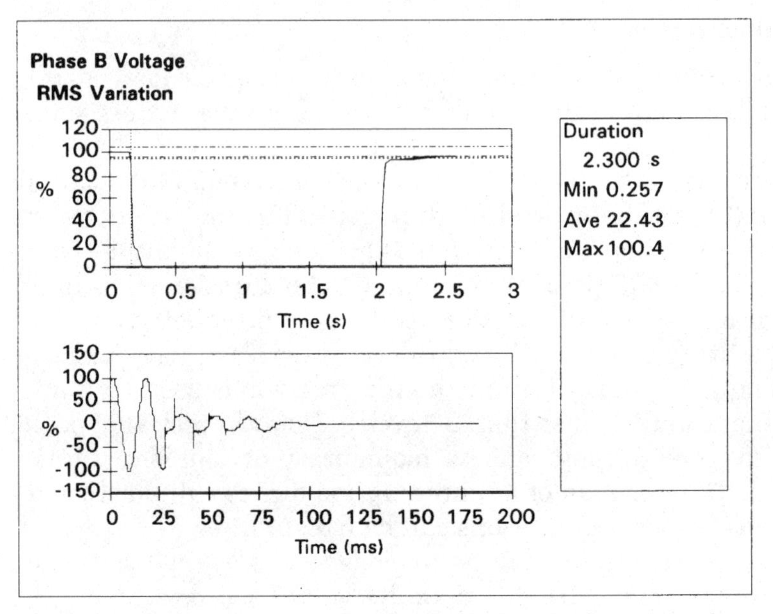

Figure 2.5 Momentary interruption due to a fault and subsequent recloser operation.

percent." When not specified otherwise, a 20 percent sag will be considered an event during which the rms voltage decreased by 20 percent to 0.8 pu. The nominal, or base, voltage level should also be specified.

Voltage sags are usually associated with system faults but can also be caused by energization of heavy loads or starting of large motors. Figure 2.6 shows a typical voltage sag that can be associated with a single-line-to-ground (SLG) fault on another feeder from the same substation. An 80 percent sag exists for about three cycles until the substation breaker is able to interrupt the fault current. Typical fault clearing times range from 3 to 30 cycles, depending on the fault current magnitude and the type of overcurrent protection.

Figure 2.7 illustrates the effect of a large motor starting. An induction motor will draw 6 to 10 times its full load current during starting. If the current magnitude is large relative to the available fault current in the system at that point, the resulting voltage sag can be significant. In this case, the voltage sags immediately to 80 percent and then gradually returns to normal in about 3 s. Note the difference in time frame between this and sags due to utility system faults.

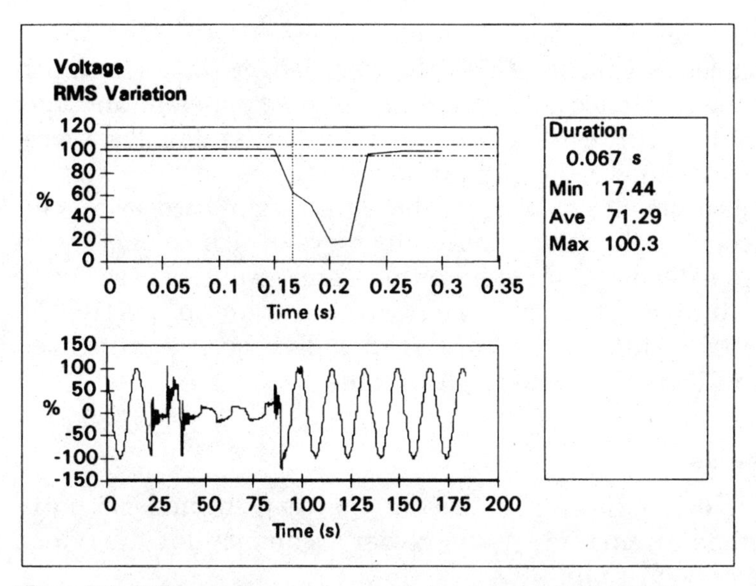

Figure 2.6 Voltage sag caused by a single-line-to-ground (SLG) fault.

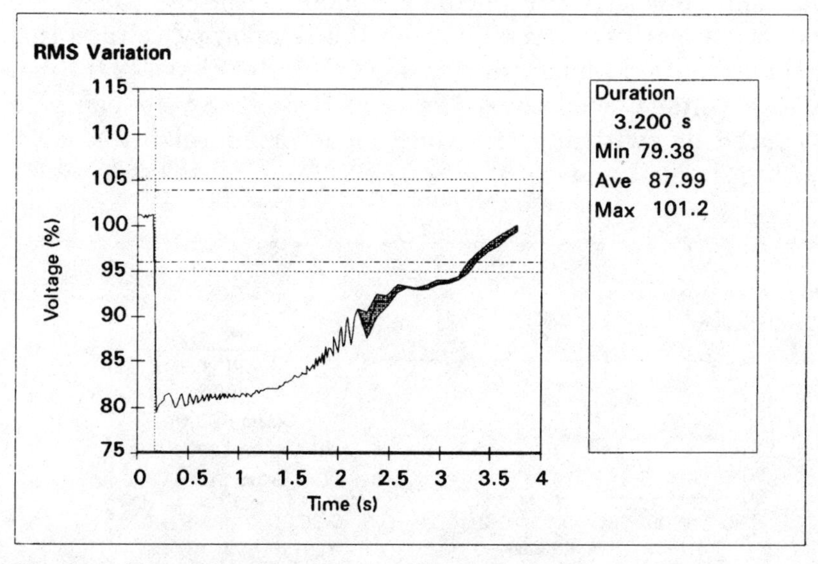

Figure 2.7 Temporary voltage sag caused by motor starting.

Until recently, the duration of sag events has not been clearly defined. Typical sag duration defined in some publications ranges from 2 ms (about one-tenth of a cycle) to a couple of minutes. Undervoltages that last less than one-half cycle cannot be characterized effectively by a change in the rms value of the fundamental frequency value. Therefore, these events are considered transients. Undervoltages that last longer than 1 min can typically be controlled by voltage regulation equipment and may be associated with causes other than system faults. Therefore, these are classified as long-duration variations.

Sag durations are subdivided here into three categories—instantaneous, momentary, and temporary—which coincide with the three categories of interruptions and swells. These durations are intended to correspond to typical utility protective device operation times as well as duration divisions recommended by international technical organizations.[5]

2.5.3 Swells

A *swell* is defined as an increase to between 1.1 and 1.8 pu in rms voltage or current at the power frequency for durations from 0.5 cycle to 1 min.

As with sags, swells are usually associated with system fault conditions, but they are not as common as voltage sags. One way that a swell can occur is from the temporary voltage rise on the unfaulted phases during a SLG fault. Figure 2.8 illustrates a voltage swell caused by a SLG fault. Swells can also be caused by switching off a large load or energizing a large capacitor bank.

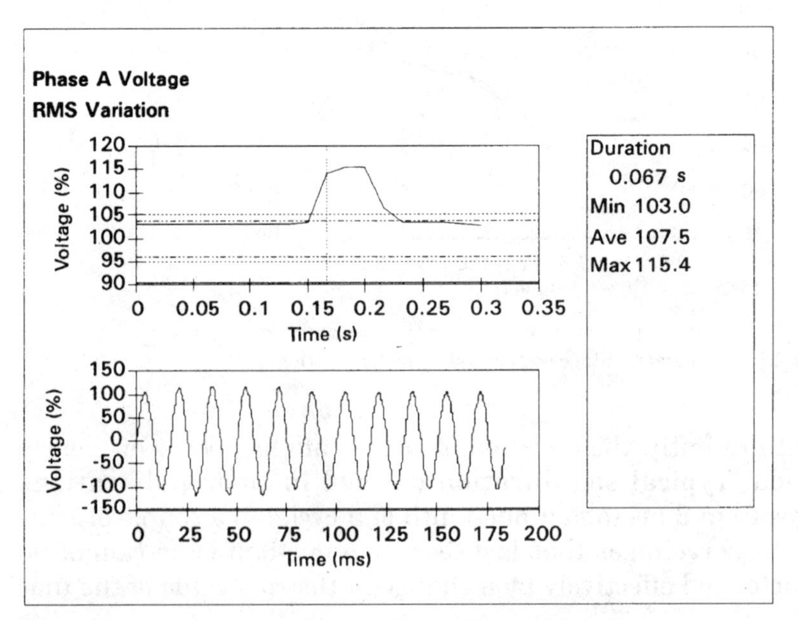

Figure 2.8 Instantaneous voltage swell caused by a SLG fault.

Swells are characterized by their magnitude (rms value) and duration. The severity of a voltage swell during a fault condition is a function of the fault location, system impedance, and grounding. On an ungrounded system, with an infinite zero-sequence impedance, the line-to-ground voltages on the ungrounded phases will be 1.73 per unit during a SLG fault condition. Close to the substation on a grounded system, there will be little or no voltage rise on the unfaulted phases because the substation transformer is usually connected delta-wye, providing a low-impedance zero-sequence path for the fault current.

Faults at different points along four-wire, multigrounded feeders will have varying degrees of voltage swells on the unfaulted phases. A 15 percent swell, like that shown in the figure, is common on U.S. utility feeders.

The term *momentary overvoltage* is used by many writers as a synonym for the term *swell*.

2.6 Voltage Imbalance

Voltage imbalance (or unbalance) is sometimes defined as the maximum deviation from the average of the three-phase voltages or currents, divided by the average of the three-phase voltages or currents, expressed in percent.

Imbalance can also be defined using symmetrical components. The ratio of either the negative- or zero-sequence component to the positive sequence component can be used to specify the percent unbalance. Figure 2.9 shows an example of these two ratios for a 1-week trend of imbalance on a residential feeder.

The primary source of voltage imbalance less than two percent is single-phase loads on a three-phase circuit. Voltage imbalance can also be the result of blown fuses in one phase of a three-phase capacitor bank. Severe voltage imbalance (greater than 5 percent) can result from single-phasing conditions.

2.7 Waveform Distortion

Waveform distortion is defined as a steady-state deviation from an ideal sine wave of power frequency principally characterized by the spectral content of the deviation.

There are five primary types of waveform distortion:

- dc offset
- Harmonics

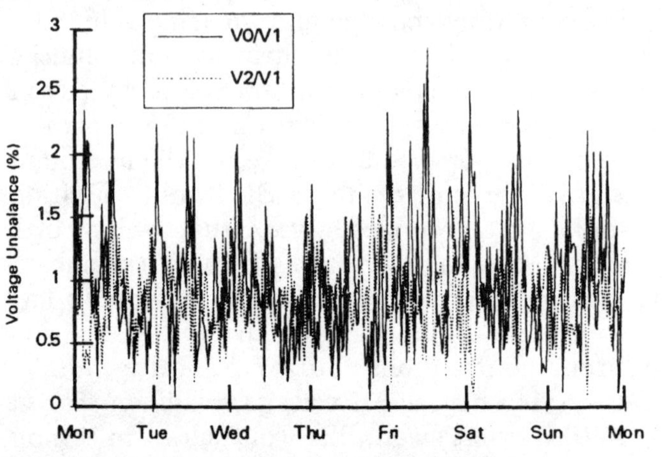

Figure 2.9 Imbalance trend for a residential feeder.

- Interharmonics
- Notching
- Noise

2.7.1 dc offset

The presence of a dc voltage or current in an ac power system is termed *dc offset.* This can occur as the result of a geomagnetic disturbance or due to the effect of half-wave rectification. Incandescent light bulb life extenders, for example, may consist of diodes that reduce the rms voltage supplied to the light bulb by half-wave rectification. Direct current in alternating current networks can have a detrimental effect by biasing transformer cores so they saturate in normal operation. This causes additional heating and loss of transformer life. Direct current may also cause the electrolytic erosion of grounding electrodes and other connectors.

2.7.2 Harmonics

Harmonics are sinusoidal voltages or currents having frequencies that are integer multiples of the frequency at which the supply system is designed to operate (termed the *fundamental* frequency; usually 50 or 60 Hz).[6] Distorted waveforms can be decomposed into a sum of the fundamental frequency and the harmonics. Harmonic distortion originates in the nonlinear characteristics of devices and loads on the power system.

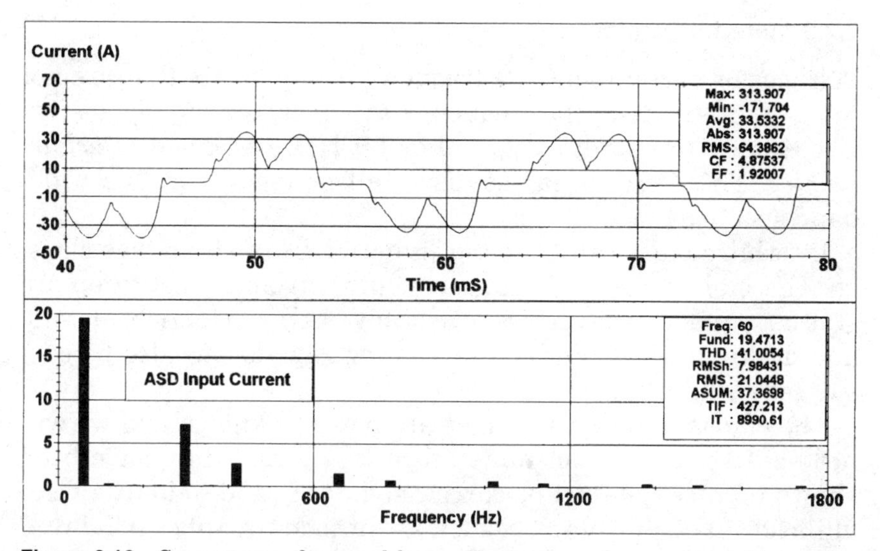

Figure 2.10 Current waveform and harmonic spectrum for an adjustable-speed-drive input current.

Harmonic distortion levels are described by the complete harmonic spectrum with magnitudes and phase angles of each individual harmonic component. It is also common to use a single quantity, the *total harmonic distortion* (*THD*), as a measure of the effective value of harmonic distortion. Figure 2.10 illustrates the waveform and harmonic spectrum for a typical adjustable-speed-drive input current. Current distortion levels can be characterized by a THD value, as described above, but this can often be misleading. For example, many adjustable-speed drives exhibit high THD values for the input current when they are operating at very light loads. This is not necessarily a significant concern because the *magnitude* of harmonic current is low, even though its relative distortion is high.

To handle this concern for characterizing harmonic currents in a consistent fashion, IEEE Standard 519-1992[9] defines another term, the *total demand distortion* (*TDD*). This term is the same as the total harmonic distortion except that the distortion is expressed as a percent of some rated load current rather than as a percent of the fundamental current magnitude. IEEE Standard 519-1992 provides guidelines for harmonic current and voltage distortion levels on distribution and transmission circuits.

2.7.3 Interharmonics

Voltages or currents having frequency components that are not integer multiples of the frequency at which the supply system is designed to operate (e.g., 50 or 60 Hz) are called *interharmonics.* They can appear as discrete frequencies or as a wideband spectrum.

Interharmonics can be found in networks of all voltage classes. The main sources of interharmonic waveform distortion are static frequency converters, cycloconverters, induction motors, and arcing devices. Power line carrier signals can also be considered as interharmonics.

The effects of interharmonics are not well known and we will not be describing them in detail in this book. They have been shown to affect power line carrier signaling, and to induce visual flicker in display devices such as cathode ray tubes (CRTs).

2.7.4 Notching

Notching is a periodic voltage disturbance caused by the normal operation of power electronics devices when current is commutated from one phase to another.

Since notching occurs continuously, it can be characterized through the harmonic spectrum of the affected voltage. However, it is generally treated as a special case. The frequency components associated with notching can be quite high and may not be readily characterized with measurement equipment normally used for harmonic analysis.

Figure 2.11 shows an example of voltage notching from a three-phase converter that produces continuous dc current. The

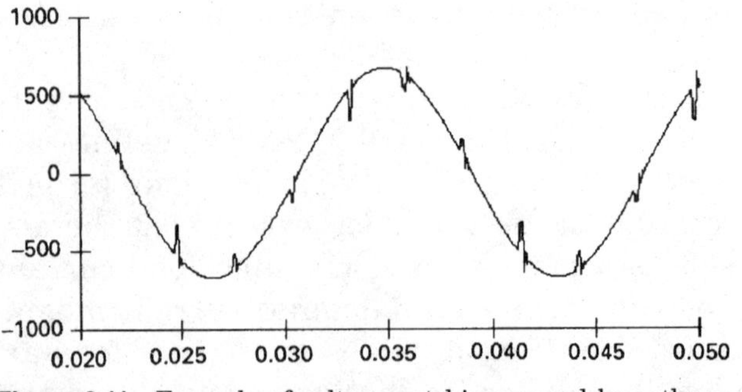

Figure 2.11 Example of voltage notching caused by a three-phase converter.

notches occur when the current commutates from one phase to another. During this period, there is a momentary short circuit between two phases pulling the voltage as close to zero as permitted by system impedances.

2.7.5 Noise

Noise is defined as unwanted electrical signals with broadband spectral content lower than 200 kHz superimposed upon the power system voltage or current in phase conductors, or found on neutral conductors or signal lines.

Noise in power systems can be caused by power electronic devices, control circuits, arcing equipment, loads with solid-state rectifiers, and switching power supplies. Noise problems are often exacerbated by improper grounding that fails to conduct noise away from the power system. Basically, noise consists of any unwanted distortion of the power signal that cannot be classified as harmonic distortion or transients. Noise disturbs electronic devices such as microcomputer and programmable controllers. The problem can be mitigated by using filters, isolation transformers, and line conditioners.

2.8 Voltage Fluctuation

Voltage fluctuations are systematic variations of the voltage envelope or a series of random voltage changes, the magnitude of which does not normally exceed the voltage ranges specified by ANSI C84.1-1982[10] of 0.9 to 1.1 pu.

IEC 1000-3-3 defines various types of voltage fluctuations. We will restrict our discussion here to IEC 1000-3-3 Type (d) voltage fluctuations, which are characterized as a series of random or continuous voltage fluctuations.

Loads which can exhibit continuous, rapid variations in the load current magnitude can cause voltage variations that are often referred to as *flicker.* The term *flicker* is derived from the impact of the voltage fluctuation on lamps such that they are perceived to flicker by the human eye. To be technically correct, voltage fluctuation is an electromagnetic phenomenon while flicker is an undesirable result of the voltage fluctuation in some loads. However, the two terms are often linked together in standards. Therefore, we will also use the common term *voltage flicker* to describe such voltage fluctuations.

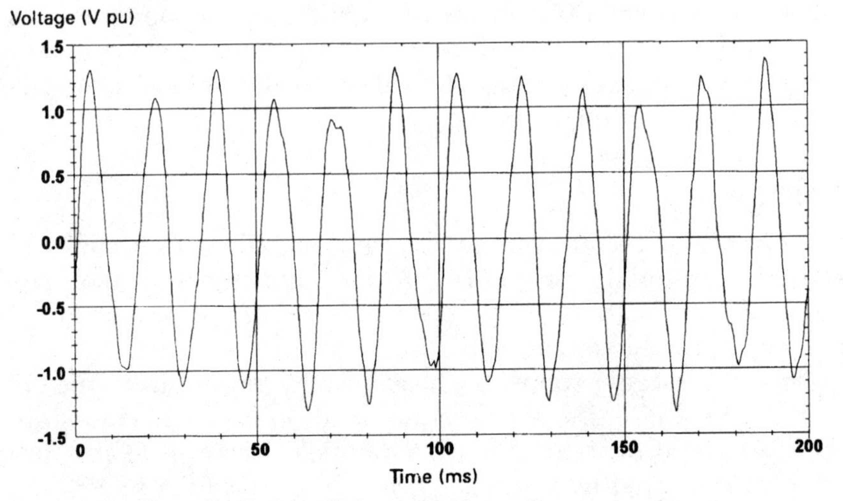

Figure 2.12 Example of voltage flicker caused by arc furnace operation.

An example of a voltage waveform which produces flicker is shown in Fig. 2.12. This is caused by an arc furnace, one of the most common causes of voltage fluctuations on utility transmission and distribution systems. The flicker signal is defined by its rms magnitude expressed as a percent of the fundamental. Voltage flicker is measured with respect to the sensitivity of the human eye. Typically, magnitudes as low as 0.5 percent can result in perceptible lamp flicker if the frequencies are in the range of 6 to 8 Hz.

2.9 Power Frequency Variations

Power frequency variations are defined as the deviation of the power system fundamental frequency from its specified nominal value (e.g., 50 or 60 Hz).

The power system frequency is directly related to the rotational speed of the generators supplying the system. There are slight variations in frequency as the dynamic balance between load and generation changes. The size of the frequency shift and its duration depends on the load characteristics and the response of the generation control system to load changes.

Frequency variations that go outside of accepted limits for normal steady-state operation of the power system can be caused by faults on the bulk power transmission system, a large block of load being disconnected, or a large source of generation going off-line.

On modern interconnected power systems, significant frequency variations are rare. Frequency variations of consequence are much more likely to occur for loads that are supplied by a generator isolated from the utility system. In such cases, governor response to abrupt load changes may not be adequate to regulate within the narrow bandwidth required by frequency-sensitive equipment.

Voltage notching can sometimes be mistaken for frequency deviation. The notches may come sufficiently close to zero to cause errors in instruments and control systems that rely on zero crossings to derive frequency or time.

2.10 Power Quality Terms

So that you will be better able to understand the material in this book, we present here the definitions of many common power quality terms. For the most part, these definitions coincide with current industry standardization efforts.[2] We have also included other relevant terms.

active filter Any of a number of sophisticated power electronic devices for eliminating harmonic distortion.

CBEMA curve A set of curves representing the withstand capabilities of computers in terms of the magnitude and duration of the voltage disturbance. Developed by the Computer Business Equipment Manufacturers Association (CBEMA), it has become a de facto standard for measuring the performance of all types of equipment and power systems, and is commonly referred to by this name.[9]

common mode voltage The noise voltage that appears equally from current-carrying conductor to ground.[2]

coupling Circuit element or elements, or network, that may be considered common to the input mesh and the output mesh and through which energy may be transferred from one to another.[8]

crest factor A value reported by many power quality monitoring instruments representing the ratio of the crest value of the measured waveform to the rms value of the waveform. For example, the crest factor of a sinusoidal wave is 1.414.

critical load Devices and equipment whose failure to operate satisfactorily jeopardizes the health or safety of personnel, and/or results in loss of function, financial loss, or damage to property deemed critical by the user.

current distortion Distortion in the ac line current. *See* distortion.
differential mode voltage The voltage between any two of a specified set of active conductors.
dip *See* sag.
distortion Any deviation from the normal sine wave for an ac quantity.
dropout A loss of equipment operation (discrete data signals) due to noise, sag, or interruption.
dropout voltage The voltage at which a device will release to its deenergized position (for this document, the voltage at which a device ceases operation).
electromagnetic compatibility The ability of a device, equipment, or system to function satisfactorily in its electromagnetic environment without introducing intolerable electromagnetic disturbances to anything in that environment.[2]
equipment grounding conductor The conductor used to connect the non-current-carrying parts of conduits, raceways, and equipment enclosures to the grounded conductor (neutral) and the grounding electrode at the service equipment (main panel) or secondary of a separately derived system (e.g., isolation transformer). See NFPA (National Fire Protection Association) 70-1990, Section 100.[7]
failure mode The effect by which failure is observed.[8]
fast tripping Refers to the common utility protective relaying practice in which the circuit breaker or line recloser operates faster than a fuse can blow. Also called fuse saving. Effective for clearing transient faults without a sustained interruption, but is somewhat controversial because industrial loads are subjected to a momentary or temporary interruption.
fault Generally refers to a short circuit on the power system.
fault, transient A short circuit on the power system usually induced by lightning, tree branches, or animals which can be cleared by momentarily interrupting the current.
flicker Impression of unsteadiness of visual sensation induced by a light stimulus whose luminance or spectral distribution fluctuates with time.[2]
frequency deviation An increase or decrease in the power frequency. The duration of a frequency deviation can be from several cycles to several hours.
frequency response In power quality usage, generally refers to the variation of impedance of the system, or a metering transducer, as a function of frequency.
fundamental (component) The component of order 1 (50 to 60 Hz) of the Fourier series of a periodic quantity.[2]

ground A conducting connection, whether intentional or accidental, by which an electric circuit or equipment is connected to the earth, or to some conducting body of relatively large extent that serves in place of the earth. Note: It is used for establishing and maintaining the potential of the earth (or of the conducting body) or approximately that potential, on conductors connected to it, and for conducting ground currents to and from earth (or the conducting body).[8]

ground electrode A conductor or group of conductors in intimate contact with the earth for the purpose of providing a connection with the ground.[7]

ground grid A system of interconnected bare conductors arranged in a pattern over a specified area and on or buried below the surface of the earth. The primary purpose of the ground grid is to provide safety for workers by limiting potential differences within its perimeter to safe levels in case of high currents which could flow if the circuit being worked became energized for any reason or if an adjacent energized circuit faulted. Metallic surface mats and gratings are sometimes utilized for the same purpose.[8] This is not necessarily the same as a signal reference grid.

ground loop A potentially detrimental loop formed when two or more points in an electrical system that are nominally at ground potential are connected by a conducting path such that either or both points are not at the same ground potential.[8]

ground window The area through which all grounding conductors, including metallic raceways, enter a specific area. It is often used in communications systems through which the building grounding system is connected to an area that would otherwise have no grounding connection.

harmonic (component) A component of order greater than one of the Fourier series of a periodic quantity.[2]

harmonic content The quantity obtained by subtracting the fundamental component from an alternating quantity.

harmonic distortion Periodic distortion of the sine wave. *See* distortion and total harmonic distortion (THD).

harmonic filter On power systems, a device for filtering one or more harmonics from the power system. Most are passive combinations of inductance, capacitance, and resistance. Newer technologies include active filters that can also address reactive power needs.

harmonic number The integral number given by the ratio of the frequency of a harmonic to the fundamental frequency.[2]

harmonic resonance A condition in which the power system is resonating near one of the major harmonics being produced by nonlinear elements in the system, thus exacerbating the harmonic distortion.

impulse A pulse that, for a given application, approximates a unit pulse or a Dirac function.[2] When used in relation to the monitoring of power quality, it is preferred to use the term *impulsive transient* in place of impulse.

impulsive transient A sudden nonpower frequency change in the steady-state condition of voltage or current that is unidirectional in polarity (primarily either positive or negative).

instantaneous When used as a modifier to quantify the duration of a short-duration variation, refers to a time range from 0.5 to 30 cycles of the power frequency.

instantaneous reclosing A term commonly applied to reclosing of a utility breaker as quickly as possible after interrupting fault current. Typical times are 18 to 30 cycles.

interharmonic (component) A frequency component of a periodic quantity that is not an integer multiple of the frequency at which the supply system is designed to operate (e.g., 50 or 60 Hz).

interruption, momentary (electric power systems) An interruption of duration limited to the period required to restore service by automatic or supervisory-controlled switching operations or by manual switching at locations where an operator is immediately available. Note: Such switching operations must be completed in a specified time not to exceed 5 min.

interruption, momentary (power quality monitoring) A type of short-duration variation. The complete loss of voltage (<0.1 pu) on one or more phase conductors for a time period between 30 cycles and 3 s.

interruption, sustained (electric power systems) Any interruption not classified as a momentary interruption.

interruption, sustained (power quality) A type of long-duration variation. The complete loss of voltage (<0.1 pu) on one or more phase conductors for a time greater than 1 min.

interruption, temporary A type of short-duration variation. The complete loss of voltage (<0.1 pu) on one or more phase conductors for a time period between 3 s and 1 min.

isolated ground An insulated equipment grounding conductor run in the same conduit or raceway as the supply conductors. This conductor is insulated from the metallic raceway and all ground points throughout its length. It originates at an isolated ground-type receptacle or equipment input terminal block and terminates at the point where neutral and ground are bonded at the power source. See NFPA 70-1990, Section 250-74, Exception #4, and Section 250-75, Exception.[7]

isolation Separation of one section of a system from undesired influences of other sections.

linear load An electrical load device which, in steady-state operation, presents an essentially constant load impedance to the power source throughout the cycle of applied voltage.

long-duration variation A variation of the rms value of the voltage from nominal voltage for a time greater than 1 min. Usually further described using a modifier indicating the magnitude of a voltage variation (e.g., undervoltage, overvoltage, or voltage interruption).

low-side surges A term coined by distribution transformer designers to describe the current surge that appears to be injected into the transformer secondary terminals upon a lightning strike to grounded conductors in the vicinity.

momentary When used as a modifier to quantify the duration of a short-duration variation, refers to a time range at the power frequency from 30 cycles to 3 s.

noise Unwanted electrical signals which produce undesirable effects in the circuits of the control systems in which they occur.[8] (For this document, "control systems" is intended to include sensitive electronic equipment in total or in part.)

nominal voltage (Vn) A nominal value assigned to a circuit or system for the purpose of conveniently designating its voltage class (such as 208/120, 480/277, or 600).[6]

nonlinear load Electrical load which draws current discontinuously or whose impedance varies throughout the cycle of the input ac voltage waveform.

normal mode voltage A voltage that appears between or among active circuit conductors.

notch A switching (or other) disturbance of the normal power voltage waveform, lasting less than a half-cycle; which is initially of opposite polarity than the waveform, and is thus subtracted from the normal waveform in terms of the peak value of the disturbance voltage. This includes complete loss of voltage for up to a half-cycle.

oscillatory transient A sudden, non–power frequency change in the steady-state condition of voltage or current that includes both positive and negative polarity value.

overvoltage When used to describe a specific type of long-duration variation, refers to a voltage having a value of at least 10 percent above the nominal voltage for a period of time greater than 1 min.

passive filter A combination of inductors, capacitors, and resistors designed to eliminate one or more harmonics. The most common variety is simply an inductor in series with a shunt capacitor, which short-circuits the major distorting harmonic component from the system.

phase shift The displacement in time of one voltage-waveform relative to other voltage-waveform(s).

power factor, displacement The power factor of the fundamental frequency components of the voltage and current wave forms.

power factor (true) The ratio of active power (watts) to apparent power (voltamperes).

pulse An abrupt variation of short duration of a physical quantity followed by a rapid return to the initial value.

reclosing The common utility practice on overhead lines of closing the breaker within a short time after clearing a fault, taking advantage of the fact that most faults are transient, or temporary.

recovery time Time interval needed for the output voltage or current to return to a value within the regulation specification after a step load or line change.[8] Also may indicate the time interval required to bring a system back to its operating condition after an interruption or dropout.

recovery voltage The voltage that occurs across the terminals of a pole of a circuit interrupting device upon interruption of the current.[8]

safety ground *See* equipment grounding conductor.

sag A decrease to between 0.1 and 0.9 pu in rms voltage or current at the power frequency for durations of 0.5 cycles to 1 min.

shield As normally applied to instrumentation cables, shield refers to a conductive sheath (usually metallic) applied, over the insulation of a conductor or conductors, for the purpose of providing means to reduce coupling between the conductors so shielded and other conductors which may be susceptible to, or which may be generating, unwanted electrostatic or electromagnetic fields (noise).

shielding Shielding is the use of a conducting and/or ferromagnetic barrier between a potentially disturbing noise source and sensitive circuitry. Shields are used to protect cables (data and power) and electronic circuits. They may be in the form of metal barriers, enclosures, or wrappings around source circuits and receiving circuits.

shielding (of utility lines) The construction of a grounded conductor or tower over the lines to intercept lightning strokes in an attempt to keep the lightning currents out of the power system.

short-duration variation A variation of the rms value of the voltage from nominal voltage for a time greater than one-half cycle of the power frequency but less than or equal to 1 min. Usually further described using a modifier indicating the magnitude of a voltage variation (e.g. sag, swell, or interruption) and possibly a modifier indicating the duration of the variation (e.g., instantaneous, momentary, or temporary).

signal reference grid (or plane) A system of conductive paths among interconnected equipment, which reduces noise-induced voltages to levels which minimize improper operation. Common configurations include grids and planes.

sustained When used to quantify the duration of a voltage interruption, refers to the time frame associated with a long duration variation (i.e., greater than 1 min.).

swell A temporary increase in the rms value of the voltage of more than 10 percent the nominal voltage, at the power frequency, for durations from 0.5 cycle to 1 min.

synchronous closing Generally used in reference to closing all three poles of a capacitor switch in synchronism with the power system to minimize transients.

temporary When used as a modifier to quantify the duration of a short-duration variation, refers to a time range from 3 s to 1 min.

total demand distortion (TDD) The ratio of the root-mean-square (rms) of the harmonic current to the rms value of the rated or maximum demand fundamental current, expressed as a percent.

total disturbance level The level of a given electromagnetic disturbance caused by the superposition of the emission of all pieces of equipment in a given system.[2]

total harmonic distortion (THD) The ratio of the root-mean-square of the harmonic content to the root-mean-square value of the fundamental quantity, expressed as a percent of the fundamental.[8]

transient Pertaining to or designating a phenomenon or a quantity which varies between two consecutive steady states during a time interval that is short compared to the time scale of interest. A transient can be a unidirectional impulse of either polarity or a damped oscillatory wave with the first peak occurring in either polarity.[2]

triplen harmonics A term frequently used to refer to the odd multiples of the third harmonic, which deserve special attention because of their natural tendency to be zero sequence.

undervoltage When used to describe a specific type of long duration variation, refers to a measured voltage having a value at least 10 percent below the nominal voltage for a period of time greater than 1 min.

voltage change A variation of the rms or peak value of a voltage between two consecutive levels sustained for definite but unspecified durations.[6]

voltage dip *See* sag.

voltage distortion Distortion of the ac line voltage. *See* distortion.

voltage fluctuation A series of voltage changes or a cyclical variation of the voltage envelope.[6]

voltage imbalance (unbalance) A condition in which the three phase voltages differ in amplitude or are displaced from their normal 120 degree phase relationship or both. Frequently expressed as the ratio of the negative sequence or zero sequence voltage to the positive sequence voltage, in percent.

voltage interruption Disappearance of the supply voltage on one or more phases. Usually qualified by an additional term indicating the duration of the interruption (e.g., momentary, temporary, or sustained.)

voltage regulation The degree of control or stability of the rms voltage at the load. Often specified in relation to other parameters, such as input-voltage changes, load changes, or temperature changes.

voltage magnification The magnification of capacitor switching oscillatory transient voltage on the primary side by capacitors on the secondary side of a transformer.

waveform distortion A steady-state deviation from an ideal sine wave of power frequency principally characterized by the spectral content of the deviation.

2.11 Ambiguous Terms

Much of the early history of the power quality movement has been marked by a fair amount of "hype" as a number of equipment vendors have jockeyed for position in the marketplace. This book attempts to apply a strong engineering interpretation of all areas of power quality and remove the hype and mystery. Marketers have created many colorful phrases to entice potential customers to buy. Unfortunately, many of these terms are ambiguous and cannot be used for technical definitions.

The following words have a varied history of usage, and some have specific definitions for other applications. The use of these words for describing power quality phenomena is discouraged unless specifically qualified by descriptive text.

Blackout	Glitch
Blink	Outage
Brownout	Interruption
Bump	Power surge
Clean ground	Raw power

Dirty ground	Spike
Clean power	Surge
Dirty power	Wink

2.12 CBEMA Curve

One of the most frequently employed displays of data to represent the power quality is the so-called CBEMA curve. A portion of the curve adapted from IEEE Standard 446[9] that we typically use in our analysis of power quality monitoring results is shown in Fig. 2.13. This curve was originally developed by the CBEMA to describe the tolerance of main frame computer equipment to the magnitude and duration of voltage variations on the power system. While many modern computers have tolerances different from this, the curve has become a standard design target for sensitive equipment to be applied on the power system and a common format for reporting power quality variation data.

The axes represent magnitude and duration of the event. Points below the envelope are presumed to cause the load to drop out due to lack of energy. Points above the envelope are presumed to cause other malfunctions such as insulation failure, overvoltage trip, and overexcitation. The upper curve is actually defined down to 0.001 cycles where it has a value of about 375 percent voltage. We typically employ the curve only from 0.1 cycles and higher due to limitations in power quality

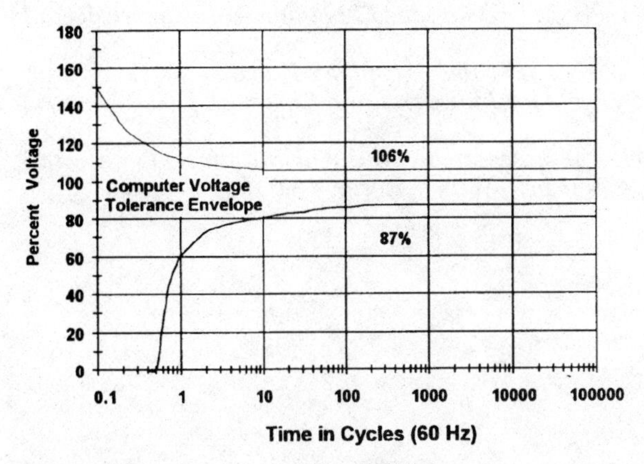

Figure 2.13 A portion of the CBEMA curve commonly used as a design target for equipment and a format for reporting power quality variation data.

monitoring instruments and differences in opinion over defining the magnitude values in the subcycle time frame.

This curve is used as a reference throughout this book to define the withstand capability of various loads and devices for protection from power quality variations. For display of power quality monitoring data, we frequently add a third axis to the plot to denote the number of events within a certain predefined cell of magnitude and duration. If restricted to just the two-dimensional view above, the plot tends to turn into a solid mass of points over time, which is not useful.

2.13 References

1. TC77WG6 (Secretary) 110-R5, *Draft Classification of Electromagnetic Environments,* January 1991.
2. IEEE P1159, *Recommended Practice on Monitoring Electric Power Quality,* Working Group on Monitoring Electrical Power Quality of SCC22—Power Quality, Draft 6, November 1994.
3. IEC 50 (161), *International Electrotechnical Vocabulary,* Chap. 161: Electromagnetic Compatibility, 1989.
4. UIE-DWG-3-92-G, *Guide to Quality of Electrical Supply for Industrial Installations—Part 1: General Introduction to Electromagnetic Compatibility (EMC), Types of Disturbances and Relevant Standards,* Advanced UIE Edition, "Disturbances" Working Group GT 2.
5. UIE-DWG-2-92-D, *UIE Guide to Measurements of Voltage Dips and Short Interruptions Occurring in Industrial Installations.*
6. IEC 1000-2-1, "Description of the Environment—Electromagnetic Environment for Low Frequency Conducted Disturbances and Signaling in Public Power Supply Systems," *Electromagnetic Compatibility (EMC)*—Part 2 Environment, Section 1, 1990.
7. ANSI/NFPA No. 70-1993, *National Electrical Code.*
8. IEEE Standard 100-1988, *IEEE Standard Dictionary of Electrical and Electronic Terms.*
9. IEEE Standard 446-1987, *IEEE Recommended Practice for Emergency and Standby Power Systems for Industrial and Commercial Applications* (IEEE Orange Book).
10. ANSI Standard C84.1-1982, *American National Standard for Electric Power Systems and Equipment—Voltage Ratings (60 Hz).*

Chapter

3

Voltage Sags and Interruptions

Voltage sags and interruptions are generally related power quality problems, so we will deal with them together to avoid repetition.

A *voltage sag* is a short-duration (typically 0.5 to 30 cycles) reduction in rms voltage caused by faults on the power system and the starting of large loads, such as motors. Momentary interruptions (typically no more than 2 to 5 s) cause a complete loss of voltage and are a common result of the actions taken by utilities to clear transient faults on their systems. Sustained interruptions of longer than 1 min are generally due to permanent faults.

In recent years, utilities have been faced with rising numbers of complaints about the quality of power due to sags and interruptions. There are a number of reasons for this, with the most important being customers with more sensitive loads in all sectors (residential, commercial, and industrial). The influx of digital computers and other types of electronic controls is at the heart of the problem. Computer controls tend to lose their memory and the processes that are being controlled also tend to be more complex, taking much more time to restart. Industries are relying more on automated equipment to achieve maximum productivity to remain competitive. Thus, an interruption has more impact than with loads common just a few years ago.

3.1 Sources of Sags and Interruptions

Voltage sags and interruptions are generally caused by faults (short circuits) on the utility system.[4] Consider a customer

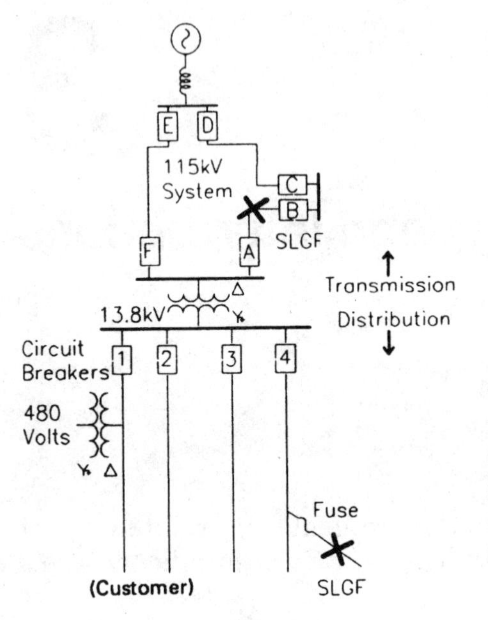

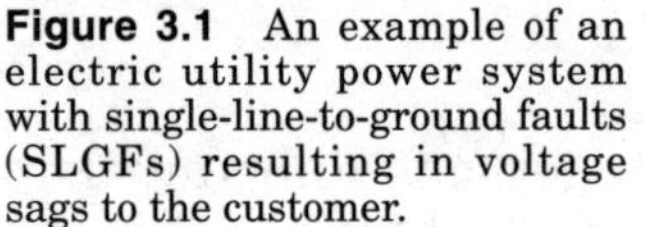
Figure 3.1 An example of an electric utility power system with single-line-to-ground faults (SLGFs) resulting in voltage sags to the customer.

that is supplied from the feeder protected by breaker 1 on the diagram shown in Fig. 3.1. If there is a fault on this feeder, the customer will experience a voltage sag during the fault followed by an interruption when the breaker opens to clear the fault. If the fault is temporary in nature, a reclosing operation on the breaker should be successful and the interruption will only be temporary. It will usually require about five or six cycles for the breaker to operate, during which time a voltage sag occurs. The breaker will remain open for a minimum of 20 cycles up to 2 to 5 s depending on utility reclosing practices. Sensitive equipment will almost surely trip during this interruption.

A much more common event would be a fault on one of the other feeders from the substation or a fault somewhere on the transmission system (see fault locations shown in Fig. 3.1). In any of these cases, the customer will experience a voltage sag during the period that the fault is actually on the system. As soon as breakers open to clear the fault, normal voltage will be restored to the customer.

Figures 3.2 and 3.3 show an interesting utility fault event recorded for an Electric Power Research Institute research project[1,7] by BMI (Basic Measuring Instruments) 8010 PQNode™ in-

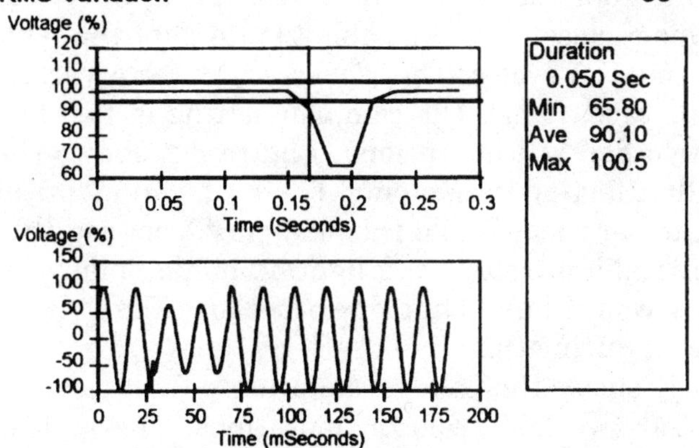

Figure 3.2 Voltage sag due to short circuit fault on a parallel electric utility feeder.

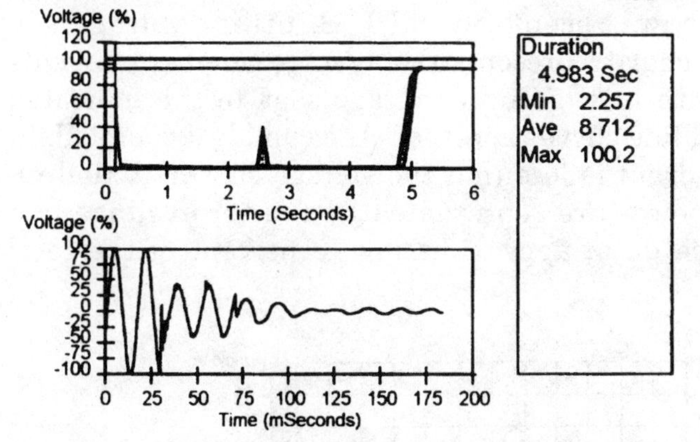

Figure 3.3 A short circuit fault event with two fast-trip operations of the utility line recloser.

struments* at two locations in the power system. The top chart in each of the figures is the rms voltage variation with time and the bottom chart is the first 175 ms of the actual waveform. Figure 3.2 shows the characteristic measured at a customer location on an unfaulted part of the feeder. Figure 3.3 shows the

*PQNode is a trademark of Basic Measuring Instruments, Santa Clara, Calif.

momentary interruption (actually two separate interruptions) observed downline from the fault. The interrupting device in this case was a line recloser that was able to interrupt the fault very quickly in about 2.5 cycles. This device can have a variety of settings. In this case, it had the common setting of two fast operations and two delayed operations. Figure 3.2 shows the brief sag for the first fast operation only. There was an identical sag for the second operation. While this is a very brief sag that is virtually unnoticeable by observing lighting blinks, many industrial processes would have shut down because the voltage sagged to 65 percent during this time.

Figure 3.3 clearly shows the voltage sag prior to fault clearing and the subsequent two fast recloser operations. The reclose time (the time the recloser was open) was a little more than 2 s, a very common time for a utility line recloser. Apparently, the fault—perhaps a tree branch—was not cleared completely by the first operation, necessitating a second operation. The system was restored after the second operation.

Figure 3.4 shows recent typical U.S. utility voltage sag data.[10] The bar chart represents the average number of events per 30-day month in which the voltage sags to the indicated range. The solid line curve represents the cumulative probability that a given event is less that the voltage shown. Actual interruptions of power are represented by the zero voltage bar. According to the data, approximately 10 percent of the total

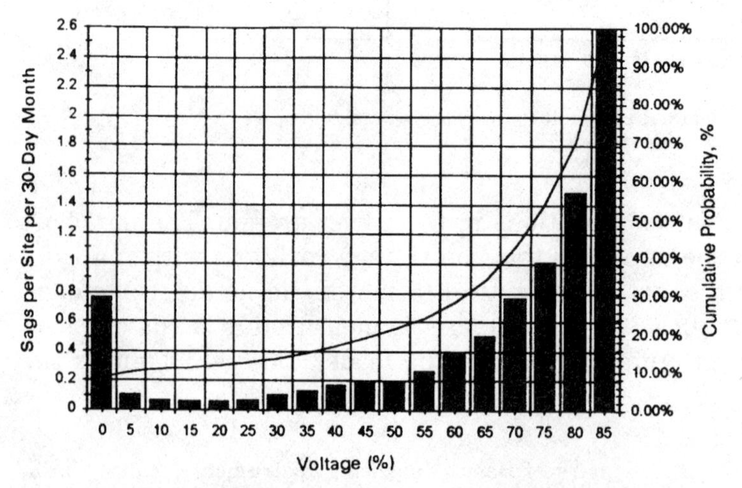

Figure 3.4 Typical U.S. electric utility sag incidence rate data for 9-month period. (*Courtesy of Electric Power Research Institute, RP 3098-1.*)[10]

events yielding a voltage of less than 90 percent are interruptions. The remainder are voltage sags of various depths caused by faults elsewhere on the system. The median sag voltage is 75 percent. While these were preliminary data, they do give a general idea about the sag and interruption frequency.

3.2 Area of Vulnerability

The concept of an "area of vulnerability" has been developed to help evaluate the likelihood of being subjected to voltage sags lower than a critical value.[5,6] Figure 3.5 shows an area of vulnerability diagram for an industrial customer supplied from a transmission system bus. The expected voltage sag performance is developed by performing short circuit simulations to determine the plant voltage as a function of fault location throughout the power system. Total circuit miles of line exposure that can affect the plant (area of vulnerability) are determined for a particular voltage sag level. The figure shows that the area of vulnerability is dependent on the sensitivity of the equipment. Contactors that drop out at 50 percent voltage will have a relatively small area of vulnerability while adjustable-speed drives (ASDs) that drop out at 90 percent voltage may be sensitive to faults over a much wider range of the transmission system.

Historical fault performance (expressed in the number of faults per year per 100 mi of line) can then be used to estimate

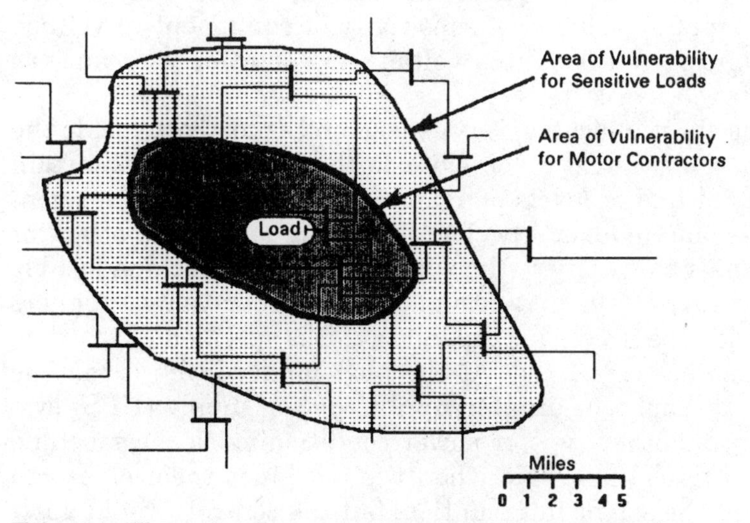

Figure 3.5 Illustration of a transmission system area of vulnerability.

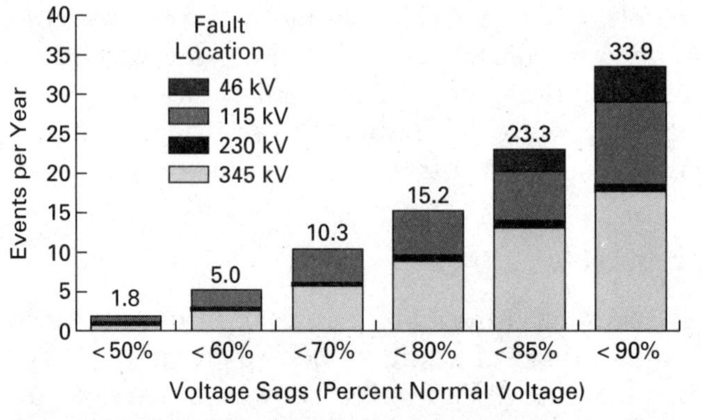

Figure 3.6 Estimated voltage sag performance at customer site as a function of faulted line voltage and voltage sag level.

the number of sags per year that can be expected below that magnitude. Finally, a chart such as the one in Fig. 3.6 can be constructed breaking down the expected voltage sags by magnitude (voltage level of faulted line) and cause. This information can be used directly by the end user to determine the need for power conditioning equipment at sensitive loads in the plant.

The same analysis may be performed for the distribution system.

3.3 Fundamental Principles of Protection

Several things can be done by the utility, end user, and equipment manufacturer to reduce the number and severity of voltage sags and to reduce the sensitivity of equipment to voltage sags. Figure 3.7 illustrates voltage sag solution alternatives and their relative costs.

As this chart indicates, it is generally less costly to tackle the problem at its lowest level, close to the load. As we entertain solutions at higher levels of available power, the solutions generally become more costly. The least-cost solution is often for the end user to specify to the supplier that the machine be able to ride through sags of a designated duration. Many suppliers can provide the necessary capability if it is specified at the time quotations are requested. At the next higher level, it may be possible to apply an uninterruptible power supply (UPS) system or some other type of power conditioning to the machine control. This is applicable when the machines themselves can withstand the sag or interruption, but the controls would automatically shut them down.

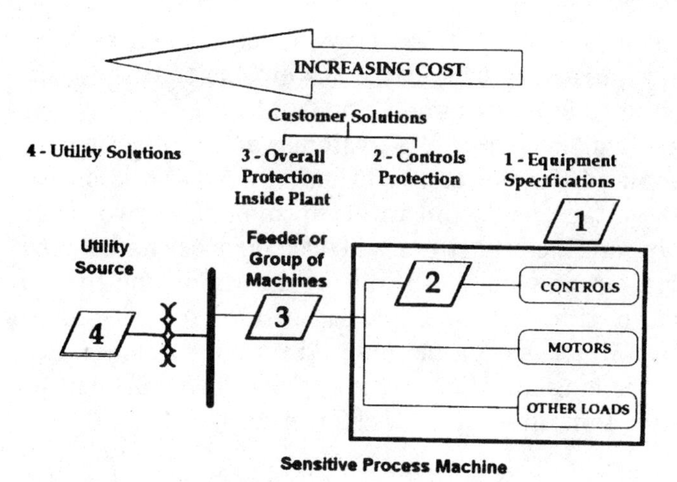

Figure 3.7 Approaches for voltage sag ride-through.

At level 3 in the figure, some sort of backup power supply with the capability to support the load for a brief period is required. Level 4 represents alterations to the utility power system to significantly reduce the number of sags and interruptions.

3.4 End-User Issues

To ride through disruptions of this variety, the load will need some kind of system that can react within about one-half cycle and provide near-normal power for a few seconds until the voltage is fully restored. This requires either a source of stored energy at the site or an alternate source of energy. These devices must either be capable of being switched very quickly or be always on-line.

Normally, because of economic constraints, protection is applied to only the most critical loads in a plant. Frequently, the critical load can be resolved to a few electronic controllers or computers, and commonly available UPS systems can be employed to handle the problem. However, much recent work has been going into supplying the whole plant for the time of the disruption. This has resulted in the development of high-energy storage devices, such as the Superconducting Storage Device (SSD)™,* and fast transfer switches that can switch to an alternate feeder within a few milliseconds. The SSD can ride through

*Trademark of Superconductivity, Inc., Madison, Wis.

interruptions of at least 2 s. Direct current loads such as telephone systems require very large UPS systems so that they can remain powered until standby generation can be started.

Ferroresonant transformers, UPS systems, and magnetic synthesizers are some of the power conditioning devices which can protect against voltage sags and interruptions. The two basic types of UPSs are on-line and standby. These devices can be used for long-duration outages up to 15 min in duration. The hybrid UPS, a variation of the standby UPS, can also be used for long-duration outages. Motor-generator sets and rotary UPSs are also being employed for long-duration interruptions. The SSD can be used for short-duration interruptions of 2 s or less.

3.4.1 Ferroresonant transformers

Ferroresonant transformers, also called constant-voltage transformers (CVTs), can handle most voltage sag conditions. CVTs are especially attractive for constant, low-power loads. Variable loads, especially with high inrush currents, present more of a problem for CVTs because of the tuned circuit on the output. Ferroresonant transformers are basically 1:1 transformers which are excited high on their saturation curves, thereby providing an output voltage which is not significantly affected by input voltage variations. A typical ferroresonant circuit is shown in Fig. 3.8. Figure 3.9 shows the voltage sag ride-through improvement of a process controller fed from a 120-VA ferroresonant transformer.

Figure 3.9 shows the marked improvement of the process controller to ride through voltage sags. The process controller can now ride through a voltage sag down to 30 percent of nomi-

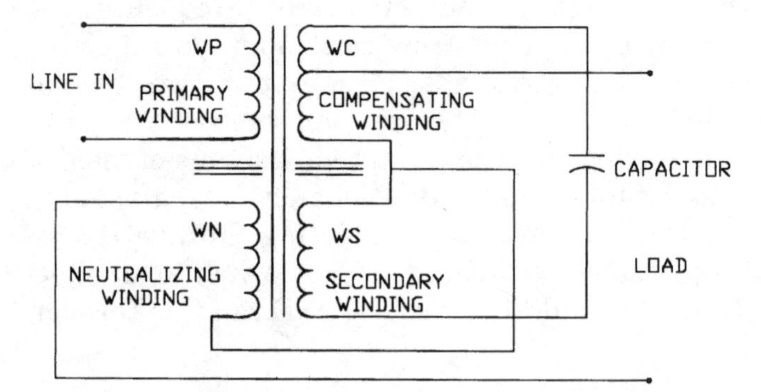

Figure 3.8 Ferroresonant constant-voltage transformer.

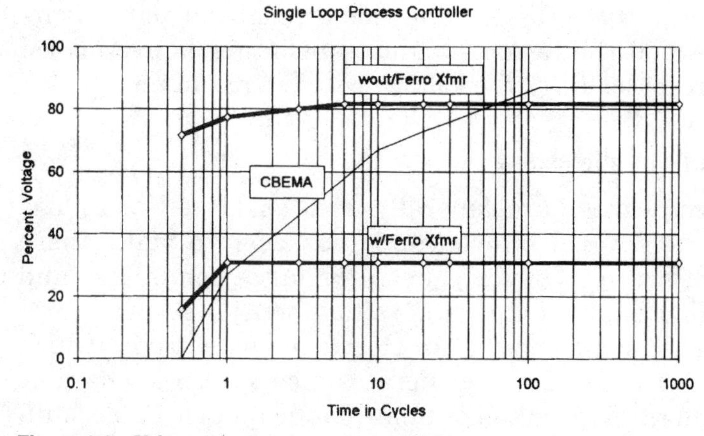

Figure 3.9 Voltage sag improvement with ferroresonant transformer.

nal with a 120-VA ferroresonant transformer, as opposed to 82 percent without one. Notice how the ride-through capability is held constant at a certain level. The reason for this is the small power requirement of the process controller, only 15 VA.

Ferroresonant transformers should be sized about four times greater than the load. Figure 3.10 shows the allowable voltage sag as a percentage of nominal voltage versus ferroresonant transformer loading, as specified by one manufacturer.

At 25 percent of loading, the allowable voltage sag is 30 percent of nominal, which means that the CVT will output over 90 percent normal voltage as long as the input voltage is above 30 percent. This is important since the plant voltage rarely falls below 30 percent of nominal during voltage sag conditions. As the load-

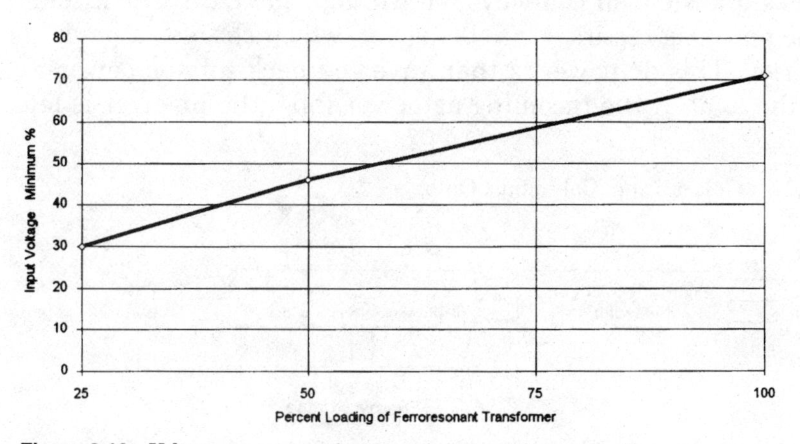

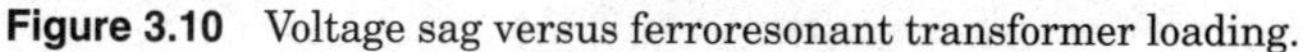

Figure 3.10 Voltage sag versus ferroresonant transformer loading.

ing is increased, the corresponding ride-through capability is reduced, and when the ferroresonant transformer is overloaded (e.g., 150 percent loading), the voltage will collapse to zero.

3.4.2 Magnetic synthesizers

Magnetic synthesizers are generally used for larger loads. The loads must be several kilovoltamperes (kVA) to make these units cost effective. They are used for large computers and other electronic equipment that is voltage sensitive.

The magnetic synthesizer is an electromagnetic device which takes incoming power and regenerates a clean, three-phase ac output waveform, regardless of input power quality. A block diagram of the process is shown in Fig. 3.11.

Energy transfer and line isolation is accomplished through the use of nonlinear chokes. This eliminates problems such as line noise. The ac output waveforms are built by combining distinct voltage pulses from saturated transformers. The waveform energy is stored in the saturated transformers and capacitors as current and voltage. This energy storage enables the output of a clean waveform with little harmonic distortion. Finally, three-phase power is supplied through a zigzag transformer. Figure 3.12 shows a magnetic synthesizer's voltage sag ride-through capability as compared to the CBEMA curve, as specified by one manufacturer.*

3.4.3 On-line UPS

Figure 3.13 shows a typical configuration of an on-line UPS. In this design, the load is always fed through the UPS. The incoming ac power is rectified into dc power, which charges a bank of batteries. This dc power is then inverted back into ac power to feed the load. If the incoming ac power fails, the inverter is fed

*Liebert Corporation, Columbus, Ohio.

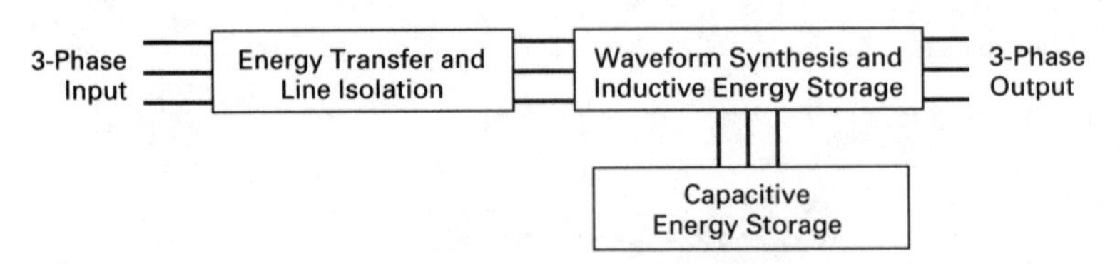

Figure 3.11 Block diagram of magnetic synthesizer.

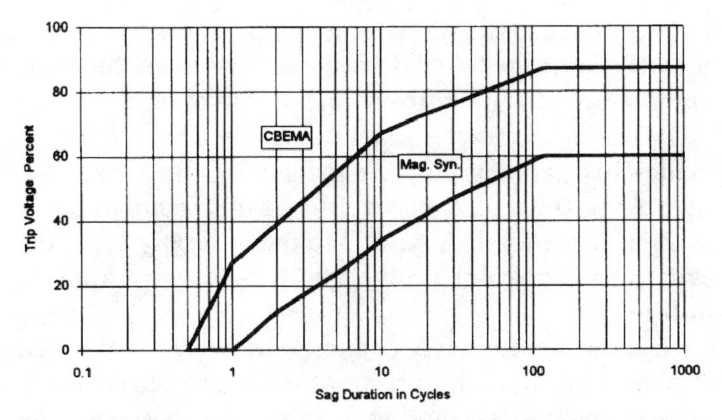

Figure 3.12 Magnetic synthesizer voltage sag ride-through capability.

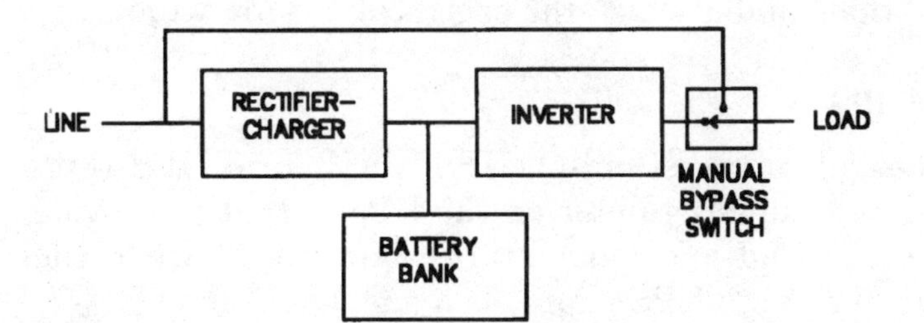

Figure 3.13 An on-line uninterruptible power supply (UPS).

from the batteries and continues to supply the load. In addition to providing ride-through for power outages, an on-line UPS provides very high isolation of the critical load from all power line disturbances. However, an on-line UPS can be quite expensive and lossy.

3.4.4 Standby UPS

A standby power supply (Fig. 3.14) is sometimes termed "off-line UPS" since the normal line power is used to power the

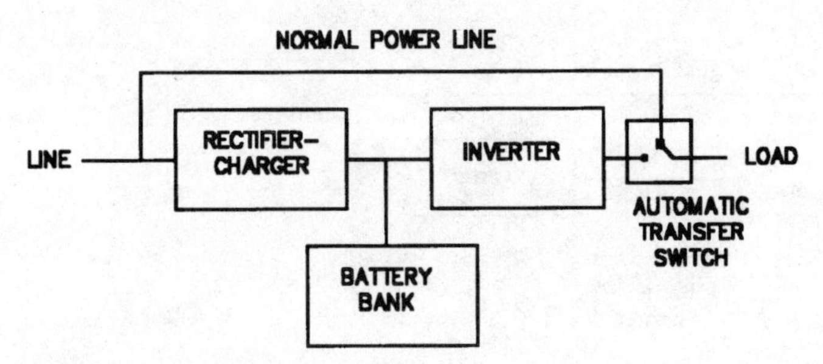

Figure 3.14 A standby UPS.

equipment until a disturbance is detected and a switch transfers the load to the battery-backed inverter. The transfer time from the normal source to the battery-backed inverter is important. The CBEMA curve shows that 8 ms is the lower limit on voltage sag ride-through for power-conscious manufacturers. Therefore a transfer time of 4 ms would ensure continuity of operation for the critical load. A standby power supply typically does not provide any transient protection or voltage regulation as does an on-line UPS.

UPS specifications include kVA capacity, dynamic and static voltage regulation, harmonic distortion of the input current and output voltage, surge protection, and noise attenuation. The specifications should indicate, or the supplier should furnish, the test conditions under which the specifications are valid.

3.4.5 Hybrid UPS

Similar in design to the stand-by UPS, the hybrid UPS (Fig. 3.15) utilizes a voltage regulator on the UPS output to provide regulation to the load and momentary ride-through when the transfer from normal to UPS supply is made.

3.4.6 Motor-generator sets

Motor-generator (M-G) sets come in a wide variety of sizes and configurations. One type of M-G set uses an electric motor-driven synchronous generator that can produce a constant 60-Hz frequency, regardless of the speed of the machine. It is able to supply a constant output by continually changing the polarity of the rotor's field poles. Thus, each revolution can have a different number of poles than the last one. Constant output is maintained as long as the rotor is spinning at speeds between 3150 and 3600 rpm. Flywheel inertia allows the generator rotor

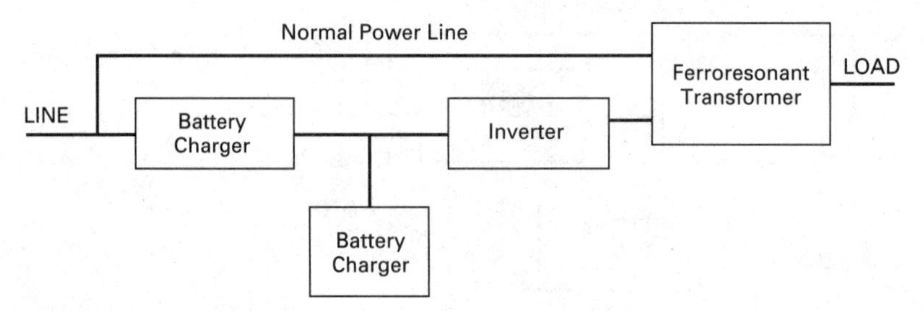

Figure 3.15 Hybrid UPS.

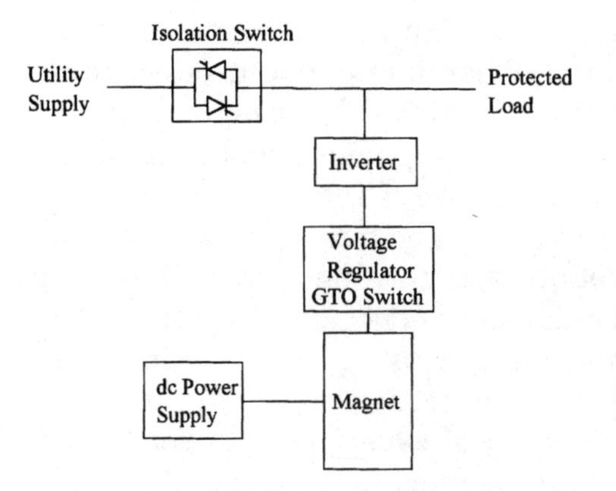

Figure 3.16 One-line diagram of a Superconducting Storage Device (SSD)™.

to keep rotating at speeds above 3150 rpm once power shuts off. The rotor weight generates enough inertia to keep it spinning fast enough to produce 60 Hz for 15 s under full load.

3.4.7 Superconducting magnetic energy storage device (SMES)

An SMES utilizes a superconducting magnet (Fig. 3.16) to store energy in the same way a UPS uses batteries to store energy.[2,12] SMES designs in the 1 to 5 MJ range are called *micro-SMESs* to distinguish them from large power sizes. The main advantage of the micro-SMES is the greatly reduced physical space needed for the magnet as compared to batteries. Fewer electrical connections are involved with a micro-SMES compared to a UPS, so the reliability should be greater and the maintenance requirements less. Initial micro-SMES designs are currently being tested in several locations with favorable results.

3.4.8 End-user equipment specifications

Another way end users can combat voltage sag problems is through their equipment procurement specifications. This essentially means keeping problem equipment out of their plant, or at least identifying ahead of time power conditioning requirements. Several ideas, outlined below, could easily be incorporated into any company's equipment procurement specifications to help alleviate problems associated with voltage sags.

1. Equipment manufacturers should have voltage sag ride-through capability curves (similar to the ones shown above),

available to their customers so that an initial evaluation of the equipment can be performed. Customers should begin to demand these types of curves to properly evaluate equipment.

2. The company procuring new equipment should establish a procedure that rates the importance of the equipment. If the equipment is critical in nature, it will be necessary to make sure adequate ride-through capability is included when the equipment is purchased. If the equipment is not important or does not cause major disruptions in manufacturing or jeopardize plant and personnel safety, voltage sag protection may not be justified.

3. Since the relative probability of experiencing a voltage sag to 70 percent or less of nominal is much less than experiencing a sag to 90 percent or less of nominal, it makes sense that if an upper limit is chosen for a ride-through capability specification, it should be somewhere in the 70 to 75 percent range. A more ideal value would be around 50 percent.

3.5 Motor-Starting Sags

Motors have the undesirable effect of drawing several times their full load current while starting. By flowing through system impedances, this large current will cause a voltage sag which may dim lights, cause contactors to drop out, and disrupt sensitive equipment. The situation is made worse by an extremely poor starting displacement factor—usually in the range of 15 to 30 percent.

The time required for the motor to accelerate to rated speed increases with the magnitude of the sag, and an excessive sag may prevent the motor from starting successfully. Motor-starting sags can persist for many seconds, as illustrated in Fig. 3.17.

3.5.1 Motor-starting methods

Energizing the motor in a single step (*full-voltage starting*) provides low cost and allows the most rapid acceleration. It is the preferred method unless the resulting voltage sag or mechanical stress is excessive.

Autotransformer starters have two autotransformers connected in open delta. Taps provide a motor voltage of 80, 65, or 50 percent of system voltage during startup. Line current and starting torque vary with the square of the voltage applied to the motor, so the 50 percent tap will deliver only 25 percent of

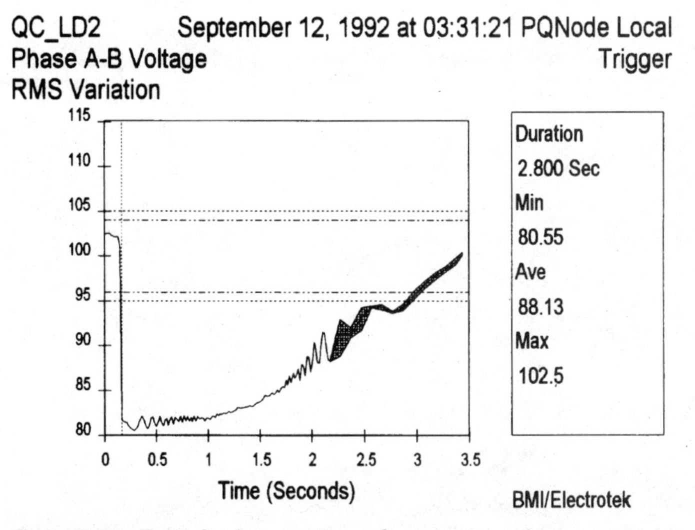

Figure 3.17 Typical motor-starting voltage sag.

the full-voltage starting current and torque. The lowest tap which will supply the required starting torque is selected.

Resistance and reactance starters initially insert an impedance in series with the motor. After a time delay, this impedance is shorted out. Starting resistors may be shorted out over several steps; starting reactors are shorted out in a single step. Line current and starting torque vary directly with the voltage applied to the motor, so for a given starting voltage, these starters draw more current from the line than with autotransformer starters, but provide higher starting torque. Reactors are typically provided with 50, 45, and 37.5 percent taps.

Part-winding starters are attractive for use with dual-rated motors (220/440 V or 230/460 V). The stator of a dual-rated motor consists of two windings connected in parallel at the lower voltage rating, or in series at the higher voltage rating. When operated with a part-winding starter at the lower voltage rating, only one winding is energized initially, limiting starting current and starting torque to 50 percent of the values seen when both windings are energized simultaneously.

Delta-wye starters connect the stator in wye for starting, then after a time delay, reconnect the windings in delta. The wye connecting reduces the starting voltage to 57 percent of the system line-line voltage; starting current and starting torque are reduced to 33 percent of their values for full-voltage start.

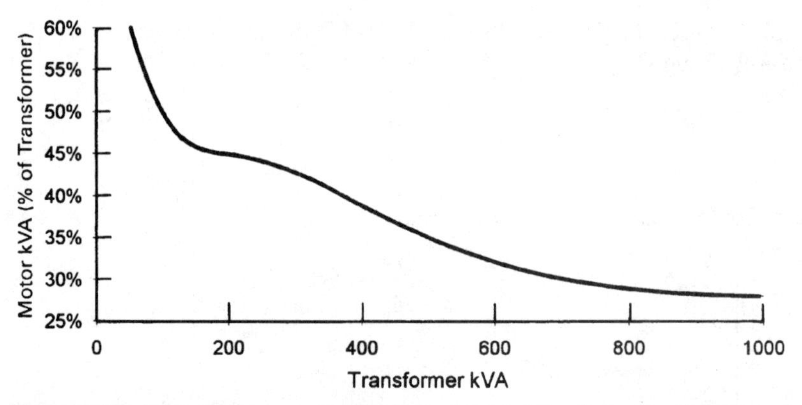

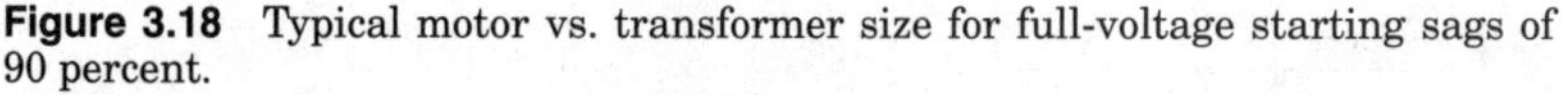
Figure 3.18 Typical motor vs. transformer size for full-voltage starting sags of 90 percent.

3.5.2 Estimating the sag severity during full-voltage starting

As shown in Fig. 3.18, starting an induction motor results in a steep dip in voltage, followed by a gradual recovery. If full-voltage starting is used, the sag voltage, in per unit of nominal system voltage is

$$V_{min}(pu) = \frac{V(pu) \cdot kVA_{SC}}{kVA_{LR} + kVA_{SC}}$$

where $V(pu)$ = actual system voltage, in per unit of nominal
kVA_{LR} = motor locked rotor kVA
kVA_{SC} = system short circuit kVA at the motor

Figure 3.18 illustrates the results of this computation for sag to 90 percent of nominal voltage, using typical system impedances and motor characteristics.

If the result is above the minimum allowable steady-state voltage for the affected equipment, then the full-voltage starting is acceptable. If not, then the sag magnitude vs. duration characteristic must be compared to the voltage tolerance envelope of the affected equipment. The required calculations are fairly complicated, and best left to a motor-starting or general transient analysis computer program. The following data will be required for the simulation:

- Parameter values for the standard induction motor equivalent circuit—R_1 (positive sequence resistance), X_1 (positive sequence reactance), R_2 (negative sequence resistance), X_2 (negative sequence reactance), and X_M (magnetizing reactance)

- Number of motor poles and rated rpm (or slip)
- WK^2 (inertia constant) values for the motor and the motor load
- Torque versus speed characteristic for the motor load

3.6 Utility System Fault-Clearing Issues

Utilities derive important benefits from activities that prevent faults. These activities not only result in improved customer satisfaction, but prevent costly damage to power system equipment.

Utilities have two basic options to continue to reduce the number and severity of faults on their system:

1. Prevent faults
2. Modify fault-clearing practices

Fault prevention activities include tree trimming, adding line arresters, insulator washing, and adding animal guards. Insulation on utility lines cannot be expected to withstand all lightning strokes. However, any line that shows a high susceptibility to lightning-induced faults should be investigated. On transmission lines, shielding can be analyzed for its effectiveness in reducing direct lightning strokes. Tower-footing resistance is an important factor in backflashovers from static wire to a phase wire. If the tower-footing resistance is high, the surge energy from a lightning stroke will not be absorbed by the ground as quickly. On distribution feeders, shielding may also be an option as is placing arresters along the line frequently. Of course, one of the main problems with distribution feeders is that storms blow tree limbs into the lines. In areas where the vegetation grows quickly, it is a formidable task to keep trees properly trimmed.

Improved fault-clearing practices may include adding line reclosers, eliminating fast tripping, adding loop schemes, and modifying feeder design. These practices may reduce the number and/or duration of momentary interruptions and voltage sags, but utility faults can never be eliminated completely.

3.6.1 Overcurrent coordination principles

It is important to understand the operation of the utility system during fault conditions. There are certain physical limitations to interrupting the fault current and restoring power. This places certain minimum requirements on loads that are expected to survive such events without disruption. Some things can also be done better on the utility system to improve the power quality than on the load side. Therefore, we will address the issues relevant to utility fault clearing with both the end user (or load equipment designer) and the utility engineer in mind.

We will address two fundamental types of faults on power systems:

1. *Transient (temporary) faults.* These are faults due to such things as overhead line flashovers that result in no permanent damage to the system insulation. Power can be restored as soon as the fault arc is extinguished. Automatic switchgear can do this within a few seconds. Some transient faults are self-clearing.
2. *Permanent faults.* These are faults due to physical damage to some element of the insulation system that requires intervention by a line crew to repair. The impact on the end user is an outage that lasts from several minutes to a few hours.

3.6.2 Relaying practices

The chief objective of utility fault-clearing practices, besides personnel safety, is the limitation of the damage to the distribution system. Therefore, the detection of faults and the clearing of the fault current must be done with the maximum possible speed without resulting in false operations for normal transient events.

The two greatest concerns for damage are typically:

1. Arcing damage to conductors and bushings
2. Through-fault damage to substation transformers, where the windings become displaced by excessive forces, resulting in a major failure

3.6.3 Fuses

The most basic overcurrent protective element on the system is a fuse. Fuses are relatively inexpensive and maintenance-free. For those reasons, they are generally used in large numbers on most utility distribution systems to protect individual transformers and feeder branches (sometimes called laterals or lateral branches). Figure 3.19 shows a typical overhead line fused cutout. Their fundamental purpose is to operate on permanent

Figure 3.19 Typical fused cutout used on electric utility systems. (*Photograph courtesy of Cooper Power Systems.*)

faults and isolate (sectionalize) the faulted section from the sound portion of the feeder. They are positioned so that the smallest practical section of the feeder is disturbed.

Fuses detect overcurrent by melting the fuse element, which generally is a metal such as tin or silver. This initiates some sort of arcing action that leads to the interruption of the current. Because it is based on a piece of metal that must accumulate heat until it reaches its melting temperature, it takes a fuse different amounts of time to operate at different levels of fault current. The time decreases as the current level increases, giving a fuse its distinctive inverse time-current characteristic (TCC), like that shown in Fig. 3.20. To achieve full-range coordination with fuses, all other overcurrent protective devices in the distribution system must adopt this same basic shape.

3.6.4 Reclosing

Because most faults on overhead lines are transient, the power can be successfully restored within several cycles after the current is interrupted. Thus, most automatic circuit breakers are designed to reclose three or four times, if needed, in rapid succession. The multiple operations are designed to permit various sectionalizing schemes to operate and to give some more persistent transient faults a second chance to clear. There are special circuit breakers for utility distribution systems called, appropriately, *reclosers,* that were designed to perform the reclosing function particularly well. The majority of faults are cleared on

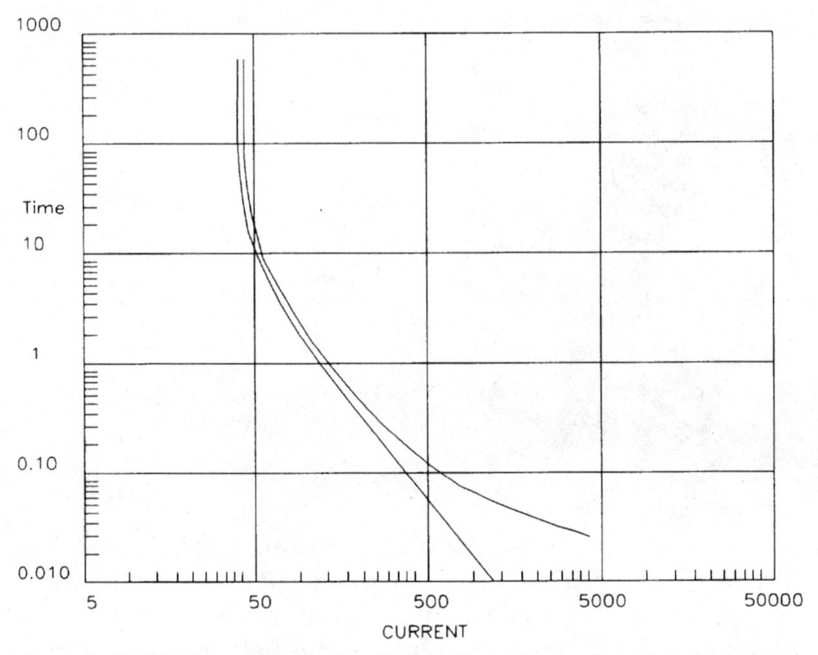

Figure 3.20 The inverse time-current characteristic of a fuse that dictates the shape of the characteristic of all other devices for series overcurrent coordination.

the first operation. Figure 3.21 shows a typical single-phase recloser and Fig. 3.22 shows a three-phase version.

Reclosing is quite prevalent in North American utility systems. Utilities in regions of low lightning incidence may reclose only once because they assume that the majority of their faults will be permanent. In lightning-prone regions, it is common to attempt to clear the fault as many as four times. Figure 3.23 illustrates the two most common sequences in use on four-shot reclosers:

1. One fast operation, three delayed
2. Two fast, two delayed

See Sec. 3.6.5 below for a more detailed explanation of fast and delayed operations.

It is generally fruitless to automatically reclose in distribution systems that are predominantly underground distribution (UD) cable unless there is a significant portion that is overhead and exposed to trees or lightning.

Figure 3.21 Typical single-phase line recloser. (*Photograph courtesy of Cooper Power Systems.*)

3.6.5 Fuse saving

Ideally, utility engineers would like to avoid blowing a fuse needlessly on transient faults because a line crew must be dispatched to change it. Line reclosers were designed specifically to help save fuses. Substation circuit breakers can use instantaneous ground relaying to accomplish the same thing. The basic idea is to have the mechanical circuit-interrupting device operate very quickly on the first operation so that it clears before any fuses downline from it have a chance to melt. When the device closes back in, power is fully restored in the majority of the cases and no human intervention is required. The only inconvenience to the customer is a slight blink. This is called the *fast* operation of the device, or the *instantaneous trip.*

If the fault is still there when the recloser or breaker recloses, there are two options:

Figure 3.22 Typical three-phase line recloser. *(Photograph courtesy of Cooper Power Systems.)*

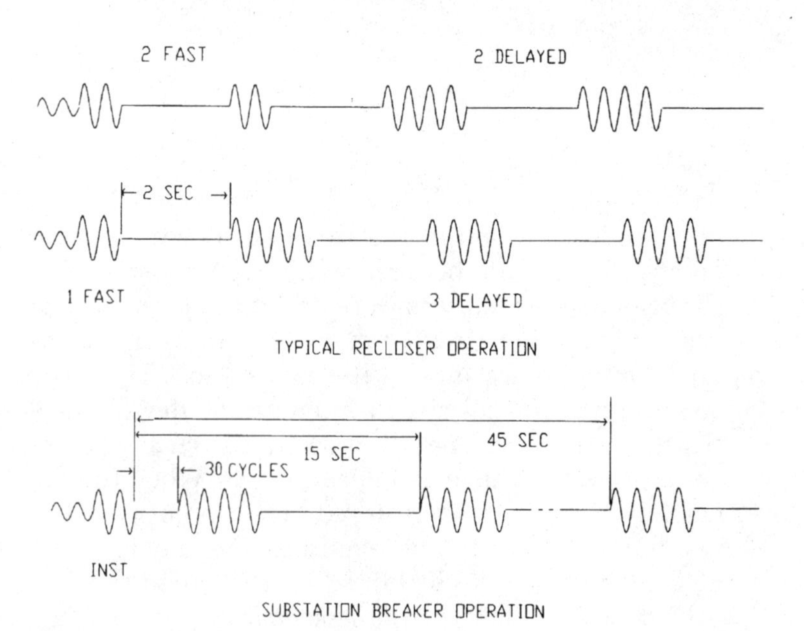

Figure 3.23 The most common reclosing sequences for line reclosers and substation breakers in use in the United States.

1. *Switch to a slow, or* delayed, *tripping characteristic.* This is frequently the only option for substation circuit breakers; they operate only once on the instantaneous trip. This philosophy assumes that the fault is now permanent and switching to a delayed operation will give a downline fuse time to operate and clear the fault by isolating the faulted section.

2. *Try a second fast operation.* This philosophy is used where experience has shown a significant percentage of transient faults need two chances to clear while saving the fuses. Some line constructions and voltage levels have a greater likelihood that a lightning-induced arc may reignite and need a second chance to clear. Also, a certain percentage of tree faults will burn free if given a second shot.

3.6.6 Reliability

The term *reliability* in the utility context usually refers to the amount of time end users are totally without power for an extended period of time (i.e., a sustained interruption). Definitions of what constitutes a sustained interruption vary among utilities in the range of 1 to 5 min. This is what many utilities refer to as an "outage." Current power quality standards efforts are leaning toward calling any interruption of power for longer than 1 min a sustained interruption (see Chap. 2). In any case, reliability is affected by the permanent faults on the system that must be repaired before service can be restored.

Of course, many industrial end users have a different view of what constitutes reliability, because even transient faults can knock their processes off-line and require several hours to get back into production. There is a movement to extend the traditional reliability indices to include momentary interruptions as well.

The traditional reliability indices for utility distribution systems are defined as follows:

SAIFI: System Average Interruption Frequency Index

$$\text{SAIFI} = \frac{\text{(No. customers interrupted)(no. of interruptions)}}{\text{Total no. customers}}$$

SAIDI: System Average Interruption Duration Index

$$\text{SAIDI} = \frac{\Sigma(\text{No. customers affected})(\text{duration of outage})}{\text{Total no. customers}}$$

CAIFI: Customer Average Interruption Frequency Index

$$\text{CAIFI} = \frac{\text{Total no. customer interruptions}}{\text{No. customers affected}}$$

CAIDI: Customer Average Interruption Duration Index

$$\text{CAIDI} = \frac{\Sigma\text{Customer interruption durations}}{\text{Total no. customer interruptions}}$$

ASAI: Average System Availability Index

$$\text{ASAI} = \frac{\text{Customer hours service availability}}{\text{Customer hours service demand}}$$

where customer hours service demand = 8760 for an entire year.

Typical target values for these indices are

Index	Target
SAIFI	1.0 h
SAIDI	1.0–1.5 h
CAIDI	1.0–1.5 h
ASAI	0.99983

These are simply design targets and actual values can, of course, vary significantly from this. Burke[8] reports the results of a survey in which the average SAIFI was 1.18, SAIDI was 76.93 min, CAIDI was 76.93 min, and ASAI was 0.999375. We have experience with utilities for which the SAIFI is usually around 0.5 and SAIDI is between 2.0 and 3.0 h. This means that the fault rate was lower than typical, at least for the bulk of the customers, but the time to repair the faults was longer. This could be common for feeders with mixed urban and rural sections. The faults are more common in the rural sections, but fewer customers are affected and it takes longer to find and repair faults.

3.6.7 Impact of eliminating fuse saving

One of the more common ways of dealing with complaints about momentary interruptions is to disable the fast-tripping, or fuse-saving, feature of the substation breaker or recloser. This avoids interrupting the entire feeder for a fault on a tap. This has been a very effective way many utilities have found to deal with *complaints* about the quality of the power. It simply minimizes the number of people inconvenienced by any single event. The penalty is that customers on the affected fused tap will suffer a sustained interruption until the fuse can be replaced, even for a transient fault. There is also an additional cost to the utility to make the trouble call, and it can have an adverse impact on the reliability indices by which some utilities are graded.

In a Utility Power Quality Practices survey conducted by Cooper Power Systems for EPRI Project RP3098-1,[7] 40 percent of the utilities responding indicated that they have responded to customer complaints by removing fast tripping. 60 percent of the investor-owned utilities (IOUs) responding indicated this practice while only 30 percent of the public power utilities (largely rural electric cooperatives) did. This may validate a widely held belief that customer sensitivity to momentary interruptions is much greater in urban areas than in rural areas.

This solution to power quality complaints does not sit well with many utilities. They would prefer the optimal technical and economical solution, which would make use of the fast-trip capability of breakers and reclosers. This not only saves operating costs, but it also improves the reliability indices. Momentary interruptions have traditionally not been reported in these indices, but only the sustained interruptions. However, when we consider the economic impact on both the end user and the utility (i.e., a value-based analysis), the utility costs can be swamped by the costs of industrial end users.[10]

If the utility has been in the practice of fuse saving, there are generally some additional costs to remove fast tripping. For example, the fused cutouts along the main feed may have to be changed for better coordination. In some cases, additional lateral fuses will have to be added so that the main feeder is better protected from faults on branches. Considering engineering time, estimates for the cost of instituting this change range from $20,000 to $40,000.

Additional operating costs to change fuses that would not have blown otherwise may be as high as $2000 per year.

While these costs may seem high, they can appear relatively small if we compare them to the costs of an end user such as a plastic bag manufacturer that can sell all the output of the plant. A single breaker operation can cost $3000 to $10,000 in lost production and extra labor charges. Thus, it is economical in the global, or value-based, sense to remove fast tripping if at least three to five interruptions (momentary and sustained combined) are eliminated each year.

The impact on the reliability indices is highly dependent on the structure of the feeder and what other sectionalizing is done. The impact can be negligible if the critical industrial load is close to the substation and the rest of the feeder can be isolated with a line recloser that does employ fast tripping. The farther out the feeder one goes with no fuse saving, the greater the impact on the reliability indices. Therefore, it is advantageous to limit the area of vulnerability to as small an area as possible and to feed sensitive customers with a high economic value of service as close to substations as possible. See the next section for more details.

Removing fast tripping will not eliminate all events that cause problems for industrial users. It will only eliminate most of the momentary interruptions. However, removing fast tripping will do nothing for voltage sags due to faults on the transmission system, other feeders, or even on fused laterals. These events can account for one-half to two-thirds of the events that disrupt industrial processes. As a rule of thumb, removing fast tripping will eliminate about one-third of the industrial process disruptions in areas where lightning-induced faults are a problem. Of course, this figure depends on the types of processes being served by the feeder.

A particular problem is when there are faults close to the substation on other feeders, or even the same feeder, but on fused taps. This causes a deep sag on all feeders connected to the bus. Two approaches that have been proposed to deal with this are (1) to install reactors on each line coming from the substation bus to limit the maximum bus sag to about 60 percent[11] and (2) to install current-limiting fuses on all branch laterals near the substation so that sags are very brief (see Sec. 3.6.12 below).

Residential end users may be quite vocal about the number of interruptions they experience, but, in most cases, there is little direct economic impact for a momentary interruption. Perhaps the biggest nuisance is resetting the dozen or so digital clocks found in most households. In fact, there may be more cases of adverse economic impact if fast tripping is eliminated. For example, homes with sump pumps may suffer more cases of flooded basements if they suffer sustained interruptions of power because the lateral fuse blew due to a temporary fault during a thunderstorm. Some utilities have taken another approach with the residential complaint problem by employing *instantaneous reclosing* while retaining the fast tripping. By getting the reclose interval down to 18 to 20 cycles, the momentary interruption is so brief that the majority of digital clocks seem to be able to ride through it. This would not be fast enough to help with industrial loads, however.

3.6.8 Increased sectionalizing

The typical utility primary distribution feeder in the United States is a radial feed from the substation breaker. In its simplest form, it consists of a main three-phase feeder with fused one- and three-phase taps as shown in Fig. 3.24.

The first step in sectionalizing the feeder further to improve overall reliability is to add a line recloser as shown in Fig. 3.25. If only reliability is of concern, one might place the recloser halfway down the feeder or at the half-load point. For power quality concerns, it might be better to locate the recloser closer to the substa-

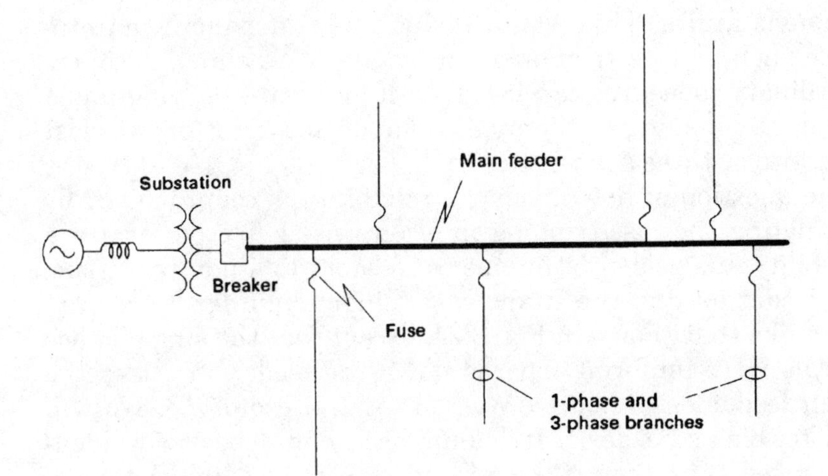

Figure 3.24 Typical main line feeder construction with fused taps.

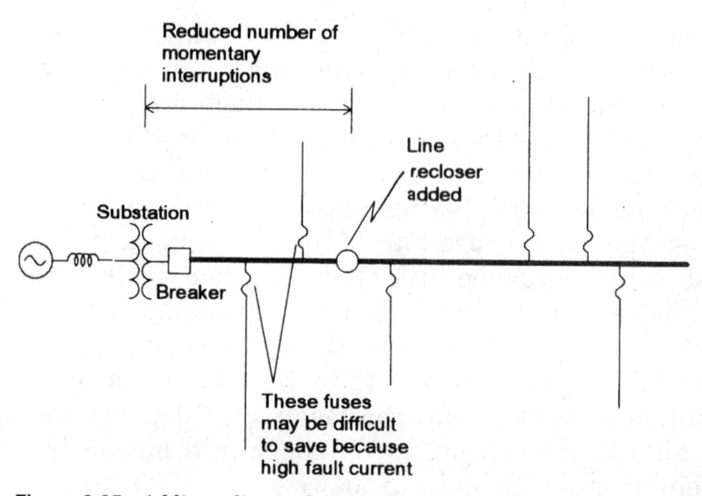

Figure 3.25 Adding a line recloser to the main feeder as the first step in sectionalizing.

tion, depending on the location of critical loads. One possible criterion is to place it at the first point where the fault current has dropped to where one can nearly always guarantee coordination with the fuses on fast tripping. Another criterion would be to place the recloser just downline from the bulk of the critical loads that are likely to complain about momentary outages.

With this concept, fast tripping can be removed from the substation breaker while only sacrificing fuse-saving on a small portion of the feeder. As pointed out above, it is often difficult to achieve fuse-saving near the substation anyway. A special effort is made to keep the first section of the main feeder free of transient faults. This would include more frequent maintenance such as tree trimming and insulator washing. Also, extraordinary measures can be taken to prevent lightning flashover, e.g., shielding or the application of line arresters at least every two or three spans.

The question of how much the reliability is compromised by eliminating the fast tripping in often raised. We performed a reliability analysis on a number of feeders to study this issue. One feeder used in the study was a single main feeder conceptually like that shown in Fig. 3.24, except that the single-phase laterals were uniformly spaced down the feeder. We used the urban feeder described by Warren[9] as the prototype. We will refer to this as Feeder 1. It is a uniform, 8-mi feeder with identical fused taps every 0.25 mi and a total of 6400 kVA load. While this may not be a realistic feeder, it is a good feeder for

TABLE 3.1 Reliability Indices Computed for Feeder 1

Case	SAIFI	SAIDI (h)	Annual fuse operations
1	0.184	0.551	1.2
2	0.299	0.666	6.0
3	0.182	0.516	1.88

study so that the general trends of certain actions can be determined. We assumed values of 0.1 faults/year/mi on the main feeder and 0.25 faults/year/mi on the fused taps, with 80 percent of the faults being transient. A uniform repair time of 3 h was assumed for permanent faults.

We first looked at the base case (Case 1 in Table 3.1), assuming that the utility was employing fuse-saving and that 100 percent of the fuses could be saved on transient faults. For Case 2, the fast tripping of the substation breaker was disabled and it was assumed that none of the tap fuses could be saved. Finally, for Case 3, we placed a three-phase recloser 1 mi from the substation and assumed that all fuses downline were saved. The resulting SAIFI and SAIDI reliability indices are shown in Table 3.1.

Typical target values for both SAIFI and SAIDI (hours) in an urban environment are 1.0. While none of these cases are particularly bad, it is apparent that removing fast tripping has a very significant negative effect on the reliability indices (compare Case 2 with Case 1). The SAIFI increases by about 60 percent. This example is a very regular, well-sectionalized feeder with a fuse on every tap, and a fuse blowing takes out less than 3 percent of the customers. For other feeder structures, the effect can be more pronounced (see below), but this serves to illustrate the point that the reliability can be expected to deteriorate when fast tripping is eliminated. The SAIDI increases only slightly. The largest change is in the number of fuse operations, which increased by a factor of 5. Thus, the utility can expect considerably more trouble calls during stormy weather.

If we subsequently add a line recloser as described for Case 3, the reliability indices and number of fuse operations return to almost the same values as the base case. In fact, the reliability indices are slightly better because of the increased sectionalizing in the line, although there are more nuisance fuse blowings in the first section than in Case 1. Thus, if we also place a line

TABLE 3.2 Reliability Indices for Feeder 2

Case	SAIFI	SAIDI (h)
1	0.43	1.28
2	1.51	2.37
3	0.47	1.29

recloser past the majority of the critical loads, eliminating fast tripping at the substation will probably not have a significant negative impact on overall reliability. Of course, this assumes that the more critical loads are close to the substation.

We studied the same three cases for another feeder that we will call Feeder 2. This feeder is, perhaps, more typical of mixed urban and rural feeders in much of the United States. Space does not allow a complete description of the topology. The main difference from Feeder 1 is that the feeder structure is more random and the sectionalizing was much more coarse with far fewer lateral fuses. The fault rate was assumed to be the same as Feeder 1. The SAIFI and SAIDI for the three cases for Feeder 2 are shown in Table 3.2. The number of fuse blowings was not computed.

For Feeder 2, many more customers are inconvenienced by each fuse blowing. Thus, the SAIFI jumps by more than a factor of 3 when fast tripping is removed. This emphasizes the need for good sectionalizing of the feeder to keep the impact on reliability at a minimum. As with Feeder 1, the Case 3 reliability indices return to nearly the same values as Case 1.

What about the power quality? Those customers in the first section of line are going to see much improved power quality as well as improved reliability. In our study of Feeder 1, the average number of interruptions, both momentary and sustained, dropped from 15 per year to a little more than 1 per year. This is a dramatic improvement! Unfortunately, the number of interruptions for the remainder of the customers—downline from the recloser—remain unchanged. What can be done about this?

The first temptation is to add another line recloser farther down on the main feeder. The customers served from the portion of the feeder between the reclosers will see an improvement. If we place the second recloser 4 mi downline on our uniform 8-mi feeder example, the average annual interruption rate drops to about 8.3. However, again, the customers at the end will see less improvement on the number of interruptions.

One can continue placing additional line reclosers in series on the main feeder and larger branch feeders to achieve even greater sectionalizing while still retaining desirable practices like fuse-saving. In this way, the portion of the feeder disturbed by a fault decreases. This will generally improve the reliability (with diminishing returns), but may not have much of an effect on the perceived power quality.

The actions that have the most effect on the number of interruptions on the portion of the feeder that is downline from the recloser are:

1. Reduce the fault rate by tree trimming, line arresters, animal guards, or other fault prevention techniques.
2. Provide more parallel branches into the service area.
3. Do not trip phases that are not involved in the fault (see Sec. 3.6.11 below).

There are at least two options for providing additional parallel paths:

1. Build more conventional feeders from the substation.
2. Use more three-phase branches from the main feeder to serve the load.

The first approach is fairly straightforward: simply build a new feeder from the substation out. This could certainly improve the reliability and power quality by simply reducing the number of customers inconvenienced by each interruption, but this may not be an economical alternative. It also may not achieve as great an improvement in the interruption rate as some of the approaches associated with the second option. Let's investigate further the second idea: using more three-phase branches off the main feeder, which has the potential of being less costly in most cases.

There are two concepts being put forward. The first involves coming out a short distance from the substation and dividing the feeder into two or three subfeeders. This could typically cut the number of interruptions by almost one-half or two-thirds, respectively, when compared to serving the same customers with a single, long main feeder. The point at which this branch occurs is a little beyond the point where it becomes practical to

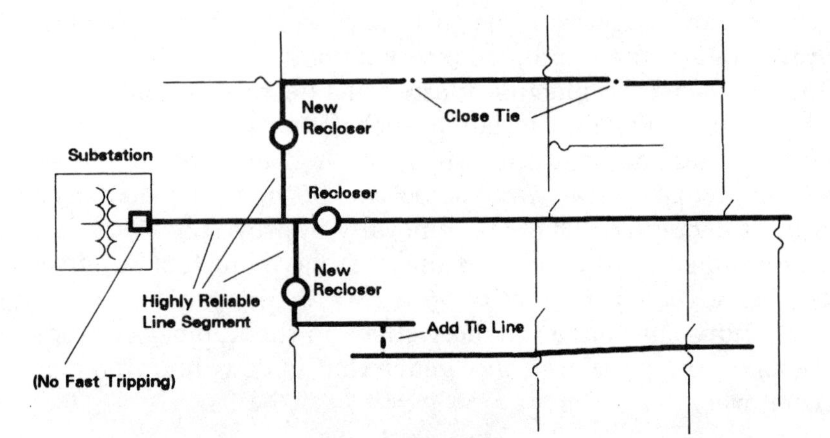

Figure 3.26 Reconfiguring a feeder with parallel subfeeders to reduce the average number of interruptions to all customers.

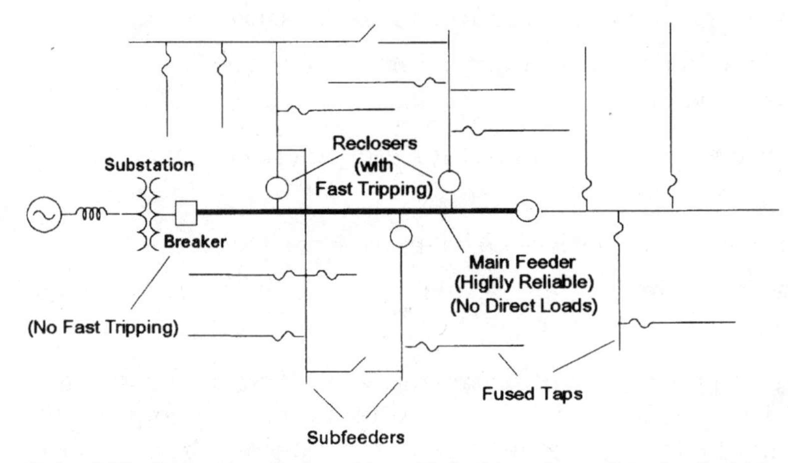

Figure 3.27 Designing a feeder with multiple three-phase subfeeds off a highly reliable main feeder.

save the vast majority of lateral fuses on temporary faults. A three-phase recloser is placed in each branch near this point. It would be wise to separate the reclosers by some distance of line to reduce the chances of sympathetic tripping, where a recloser on the unfaulted branch trips as a result of the transient currents related to the fault. Figure 3.26 depicts how this principle might be put into practice on an existing feeder with minimal rebuilding, assuming the existence of three-phase feeders of sufficient conductor size in the locations indicated.

The second proposal, as depicted in Fig. 3.27, is to first build a highly reliable main feeder that extends a significant distance into the service area. Very few loads are actually served directly off this main feeder. Instead, the loads are served off three-phase branch feeders that are tapped off the main feeder periodically. A three-phase line recloser is used at the head of each branch feeder. Of course, there is no fast tripping at the substation, so the main feeder remains as free of interruptions as possible. Special efforts will be made to prevent faults on this part of the feeder.

Essentially, the main feeder becomes an extension of the substation bus that is permitted by design to have a few more faults over its lifetime than the bus. And the branch feeders are analogous to having separate feeds to each part of the service area directly from the substation, but, hopefully, with less cost.

Whether either of these ideas is suitable for a particular utility is dependent on many factors, including terrain, load density, load distribution, and past construction practices. These ideas are presented here simply as alternatives to consider for achieving overall lower average interruption rates than is possible by stacking fault interrupters in series. Although these practices may not become widespread, they may be very useful for dealing with particularly difficult power quality complaint problems stemming from excessive interruptions.

3.6.9 Midline or tap reclosers

Despite responding to complaints by removing fast tripping, about 40 percent of the utilities surveyed indicated they were interested in adding *more* line reclosers to improve customer's perception of power quality. This would accomplish greater sectionalizing of the feeder and, perhaps, permit the use of fuse-saving practices on the bulk of the feeder again. This practice is very effective if the whole feeder is being interrupted for faults that are largely constrained to a particular region.

By putting the recloser farther out on the feeder, it will attempt to clear the fault first so that the number of customers inconvenienced by a "blink" is reduced. If it is also necessary to eliminate fast tripping on the substation breaker, only a smaller portion of the feeder nearer the substation is threatened with the possibility of having a fuse blow on a transient fault as explained above. This is not much different than the normal case because of the difficulty in preventing fuse blowing in the high fault current regions near the substation anyway.

A few utilities have actually done the opposite of this and removed line reclosers in response to complaints about momentary interruptions. Perhaps a section of the feeder ran through heavily wooded areas causing frequent operations of the recloser, or the device was responding to high ground currents due to harmonics or load imbalance, causing false trips. Whatever the reason, this is an unusual practice and is counter to the direction most utilities seem to be taking. The main question at this point does not seem to be that more line reclosers are needed, but how to go about applying them to achieve the dual goal of increased power quality and reliability of service.

3.6.10 Instantaneous reclosing

Instantaneous reclosing is the practice of reclosing within 20 to 30 cycles after interrupting the fault, generally only on the first operation. The capability of breakers and reclosers to do this has been around for several decades and some utilities use it as standard practice, particularly on substation breakers. However, the practice is not universally accepted. Many utilities reclose no faster than 2 s (the standard reclosing interval on a hydraulic recloser) and some wait even longer.

After it was observed that many digital clocks can successfully ride through a 0.5-s interruption, some utilities began to experiment with using instantaneous reclosing while retaining the fast tripping to save fuses. One utility trying this on 12-kV feeders has reported no significant increase in the number of breaker and recloser operations and that the number of complaints has diminished.[10] Therefore, it is something that other utilities that are not already using instantaneous tripping might consider, with the caution that the same experience may not be achieved at higher voltage levels and with certain line designs.

Instantaneous reclosing has had a bad reputation in some circles. One risk is that there will be insufficient time for the arc products to disperse and the fault will not clear. Some utilities have had this experience with particular voltage levels and line constructions. When this happens, substation transformers are subjected to repeated through-faults unnecessarily. This could result in increased failures of the transformers. However, if there is no indication that instantaneous reclosing is causing increased breaker operations, it should be safe to use it.

Another concern is that very high torques will be generated in rotating machines upon reclosing. Some consultants in cogeneration are quite concerned with this because 20 to 30 cycles may not be sufficient time to guarantee that the generator's protective relaying will detect a problem on the utility side. They would prefer to allow several seconds so that there is less chance the utility will reclose into the cogeneration out of synchronism. One way the utility can help prevent such an occurrence is to use a common recloser accessory that blocks reclosing when there is voltage present on the load side.

3.6.11 Single-phase tripping

Most of the three-phase breakers and reclosers on the utility distribution system operate all three phases simultaneously. One approach that has been suggested to minimize the exposure of customers to momentary outages is to trip only the faulted phase or phases. This would automatically reduce the exposure by two-thirds for most faults. The main problem with this is that it is possible to damage some three-phase loads if they are single-phased for any length of time. Thus, it is generally considered undesirable to use single-phase reclosers on three-phase branches with significant three-phase loads. Of course, this is done quite commonly when only one-phase loads are being served.

What is needed for three-phase loads is a three-phase breaker, or recloser, that is capable of operating each phase independently until it is determined that the fault is permanent. Then, to prevent single-phasing of three-phase loads, the device must lock out all three phases. A few such devices are available, but they would have to be available in a wider range of ratings for their application to become sufficiently widespread to have an impact on overall power quality.

3.6.12 Current-limiting fuses

Current-limiting fuses are often used in electrical equipment where the fault current is very high and an internal fault could result in a catastrophic failure. Since they are more expensive than conventional expulsion links, their application is generally limited to locations where the fault current is in excess of 2000 to 3000 A. Figure 3.28 shows examples of current-limiting fuses. There are various designs, but the basic configuration is

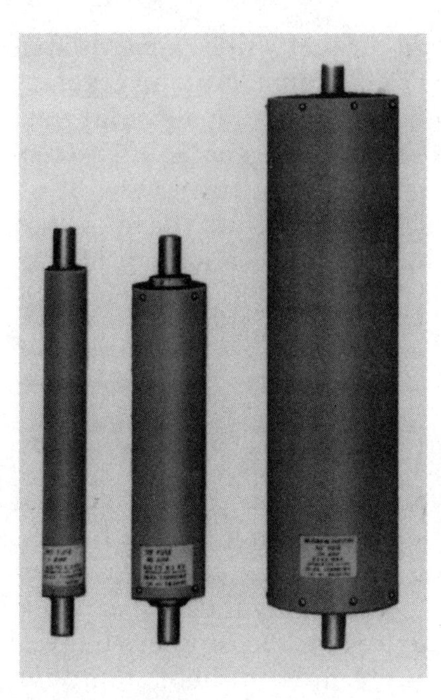

Figure 3.28 Typical current-limiting fuses used in electric utility applications. (*Photograph courtesy of Cooper Power Systems.*)

that of a thin ribbon element or wire wound around a form and encased in sand. The element melts in many places simultaneously and, with the aid of the melting sand, very quickly builds up a voltage drop that opposes the flow of current. The current is forced to zero in about 0.25 cycle.

Current-limiting fuses have the beneficial side effect with respect to power quality that the voltage sag resulting from the fault is very brief. Figure 3.29 shows typical voltage and current waveforms from a current-limiting fuse operation during a single-line-to-ground fault. The voltage drops immediately due to the fault, but shortly recovers and overshoots to 120.9 percent as the peak arc voltage develops to cut off current flow. Note that the fault current waveform in the figure is clipped. The voltage sag is so short that it is very unlikely that industrial processes will be adversely affected. Therefore, one proposed practice is to install current-limiting fuses on each lateral branch in the high fault current region near the substation to reduce the number of sags that affect industrial processes.

When current-limiting fuses were first installed on utility systems in great numbers, there was the fear that the peak arc voltage transient, which exceeds system voltage, would cause damage to arresters and to insulation in the system. This has not proven to be a significant problem. The overvoltage is on

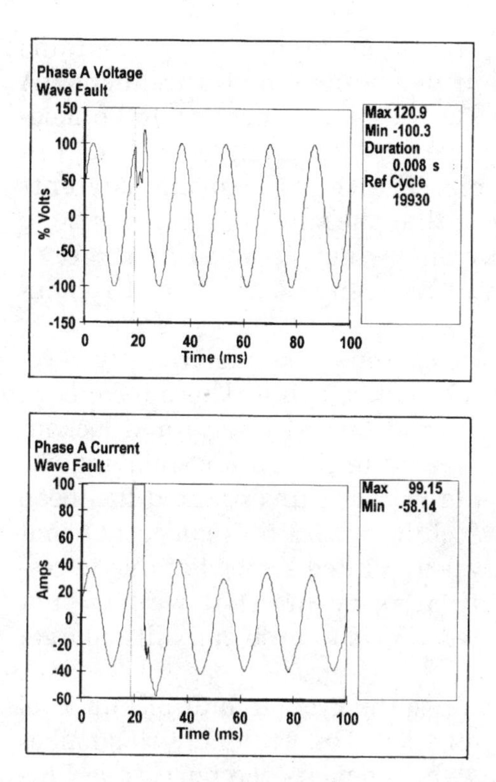

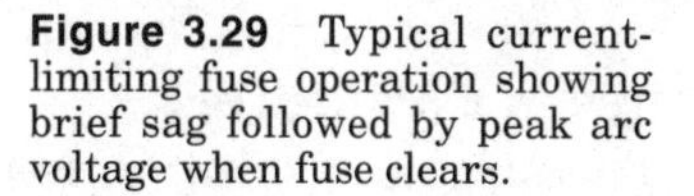
Figure 3.29 Typical current-limiting fuse operation showing brief sag followed by peak arc voltage when fuse clears.

the same order as capacitor-switching transient overvoltages, which occur several times a day on most utility systems without serious consequences.

3.6.13 Adaptive relaying

Adaptive relaying is the practice of changing the relaying characteristics of the overcurrent protective device to suit the present system conditions.

The most common thing that is currently being done with adaptive relaying is the enabling and disabling of fast tripping in response to weather conditions. This is generally done through radio or hardwire link to the utility control center. It could also be done with local devices that have the ability to detect the presence of lightning or rain.

3.6.14 Ignoring third-harmonic currents

The levels of third-harmonic currents have been increasing due to the increase in the numbers of computers and other types of electronic loads on the system. The residual current (sum of the three phase currents) on many feeders contains as much third harmonic as it does fundamental frequency. A common case is to find each of the phase currents to be moderately distorted with a THD of 7 to 8 percent, consisting primarily of the third harmonic. The third-harmonic currents sum directly in the neutral so that the third harmonic is 20 to 25 percent of the phase current, which is often as large, or larger, than the fundamental frequency current in the neutral. (See Sec. 5.6.)

Because the third-harmonic current is predominantly zero-sequence, it affects the ground-fault relaying. There have been a few incidents where there have been false trips and lockout due to excessive harmonic currents in the ground-relaying circuit. At least one of the events we have investigated has been correlated with capacitor switching where it is suspected that the third-harmonic current was amplified somewhat due to resonance. There may be many more events that we have not heard about and it is expected that the problem will only get worse in the future.

The simplest solution is to raise the ground-fault pickup level when operating procedures will allow. Unfortunately, this makes fault detection less sensitive, which defeats the purpose of having ground relaying, and some utilities are restrained by standards from raising the ground trip level. It has been observed that if the third harmonic could be filtered out, it might be possible to set the ground relaying to be *more* sensitive. The third-harmonic current is almost entirely a function of load and is not a component of fault current. When a fault occurs, the current seen by the relaying is predominantly sinusoidal. Therefore, it is not necessary for the relaying to be able to monitor the third harmonic for fault detection.

The first relays were electromagnetic devices that basically responded to the effective (rms) value of the current. Thus, for years, it has been common practice to design analog electronic relays to duplicate that response and digital relays have also generally included the significant lower harmonics. In retrospect, it would have been better if the third harmonic had been ignored for ground relays.

There is still a valid reason for monitoring the third harmonic in phase relaying because phase relaying is used to detect overload as well as faults. Overload evaluation is generally an rms function.

3.6.15 Utility fault prevention

One sure way to eliminate complaints about utility fault clearing operations is to eliminate faults altogether. Of course, there will always be some faults, but there are many things that can be done to dramatically reduce the incidence of faults.

Overhead line maintenance

Tree trimming. This is one of the more effective methods of reducing the number of faults on overhead lines. It is a necessity, although some may complain about the environmental and aesthetic impact.

Insulator washing. Like tree trimming in wooded regions, insulator washing is necessary in coastal and dusty regions. Otherwise, there will be numerous insulator flashovers for even a mild rain storm without lightning.

Shield wires. Shield wires for lightning are common for utility transmission systems. They are generally not applied on distribution feeders except where lines have an unusually high incidence of lightning strikes. Some utilities construct their feeders with the neutral on top, perhaps even extending the pole, to provide shielding. No shielding is perfect.

Improving pole grounds. Several utilities have reported doing this to improve the power quality with respect to faults. However, we are not certain of all the reasons for doing this. Perhaps, it makes the faults easier to detect. If shielding is employed, this will reduce the backflashover rate. If not, it would not seem that this would provide any benefit with respect to lightning unless combined with line arrester applications (see below).

Modified conductor spacing. Employing a different line spacing can sometimes increase the withstand to flashover or the susceptibility to getting trees in the line.

Tree wire (insulated/covered conductor). In areas where tree trimming is not practical, insulated or covered conductor can reduce the likelihood of tree-induced faults.

UD cables. Fault prevention techniques in UD cables are generally related to preserving the insulation against voltage surges. The insulation degrades significantly as it ages, requiring increasing efforts to keep the cable sound. This generally involves arrester protection schemes to divert lightning surges coming from the overhead system, although there are some efforts to restore insulation levels through injecting fluids into the cable.

Since nearly all cable faults are permanent, the power quality issue is more one of finding the fault location quickly so that the cable can be manually sectionalized and repaired. Fault location devices available for that purpose are addressed in the next section below.

Line arresters. To prevent overhead line faults, one must either raise the insulation level of the line, prevent lightning from striking the line, or prevent the voltage from exceeding the insulation level. The third idea is becoming more popular with improving surge arrester designs. To accomplish this goal, surge arresters are placed every two or three poles along the feeder as well as on distribution transformers. Some utilities place them on all three phases while other utilities place them only on the phase most likely to be struck by lightning. To support some of the recent ideas about improving power quality, or providing custom power with super-reliable main feeders, it will be necessary to put arresters on every phase of every pole.

Presently, applying line arresters in addition to the normal arrester at transformer locations is done only on line sections with a history of numerous lightning-induced faults. However, recently some utilities have claimed that line arresters are not only more effective than shielding, but more economical.[13]

Some sections of urban and suburban feeders will naturally approach the goal of an arrester every two or three poles because the density of load requires the installation of a distribution transformer at least that frequently. Each transformer will normally have a primary arrester in lightning-prone regions.

3.6.16 Fault locating

Finding faults quickly is an important aspect of reliability and the quality of power.

Faulted circuit indicators. Finding cable faults is often a challenge. The cables are underground and it is generally impossible to see the fault, although occasionally there will be a physical display. To expedite locating the fault, many utilities use "faulted circuit indicators," or simply "fault indicators," to locate the faulted section more quickly. These are devices that flip a target indicator when the current exceeds a particular level. The idea is to put one at each padmount transformer and the last one to have flipped will be just before the faulted section.

There are two main schools of thought on the selection of ratings of faulted circuit indicators. The more traditional school says to choose a rating that is two to three times the maximum expected load on the cable. This results in a fairly sensitive fault detection capability.

The opposing school says that this is too sensitive and is the reason that many fault indicators give false indication. False indication delays the location of the fault and contributes to degraded reliability and power quality. The reason given for the false indication is that the energy stored in the cable generates sufficient current to trip the indicator when the fault occurs. Thus, a few indicators downline from the fault may also show the fault. The solution to this problem is to apply the indicator with a rating based on the maximum fault current available rather than the maximum load current. This is based on the assumption that most cable faults quickly develop into bolted faults. Therefore the rating is selected allowing a margin of 10 to 20 percent.

Fault indicators must be reset before the next fault event. Some must be reset manually, while others have one of a number of techniques for detecting, or assuming, the restoration of power and resetting automatically. Some of the techniques include test point reset, low-voltage reset, current reset, electrostatic reset, and time reset.

Locating cable faults without fault indicators. Without fault indicators, the utility must rely on more manual techniques for finding the location of a fault. There are a large number of different types of fault-locating techniques and a detailed description of each is beyond the scope of this report. Some of the general classes of methods are:

1. *Thumping.* This is a common practice with numerous minor variations. The basic technique is to place a dc voltage on the cable that is sufficient to cause the fault to be reestablished and then try to detect by sight, sound, or feel the physical display from the fault. One common way to do this is with a capacitor bank that can store enough energy to generate a sufficiently loud noise. Those standing on the ground on top of the fault can feel and hear the "thump" from the discharge. Some

combine this with cable radar techniques to confirm estimates of distance. Many are concerned with the potential damage to the sound portion of the cable due to thumping techniques.

2. *Cable radar and other pulse methods.* These techniques make use of traveling-wave theory to produce estimates of the distance to the fault. The wave velocity on the cable is known. Therefore, if an impulse is injected into the cable, the time for the reflection to return will be proportional to the length of the cable to the fault. An open circuit will reflect the voltage wave back positively while a short circuit will reflect back negatively. The impulse current will do the opposite. If the routing of the cable is known, the fault location can be found simply by measuring along the route. It can be confirmed and fine-tuned by thumping the cable. On some systems, there are several taps off the cable. The distance to the fault is only part of the story; one has to determine which branch it is on. This can be a very difficult problem that is still a major obstacle to rapidly locating a cable fault.

3. *Tone.* A tone system injects a high-frequency signal on the cable and the route of the cable can be followed by a special receiver. This technique is sometimes used to trace cable routes while energized, but is also useful for fault location because the tone will disappear beyond the fault location.

4. *Fault chasing with a fuse.* The cable is manually sectionalized and then each section is reenergized until a fuse blows. The faulted section is determined by process of elimination or by observing the physical display from the fault. Because of the element of danger and the possibility of damaging cable components, some utilities strongly discourage this practice. Others require the use of small current-limiting fuses, which minimize the amount of energy permitted into the fault. This can be an expensive and time-consuming procedure that some consider to be the least effective of fault-locating methods; it is to be used only when nothing else is available.

3.7 References

1. J. Lamoree, J. C. Smith, P. Vinett, T. Duffy, and M. Klein, "The Impact of Voltage Sags on Industrial Plant Loads," in *Proceedings of the First International Conference on Power Quality* (*PQA '91*), Paris.
2. P. Vinett, R. Temple, J. Lamoree, C. De Winkel, and E. Kostecki, "Application of a Superconducting Magnetic Energy Storage Device to Improve Facility Power Quality," in *Proceedings of the Second International Conference on Power Quality: End-use Applications and Perspectives* (*PQA '92*), Atlanta, September 28–30, 1992.

3. G. Beam, E. G Dolack, C. J. Melhorn, V. Misiewicz, and M. Samotyj, "Power Quality Case Studies—Voltage Sags: The Impact on the Utility and Industrial Customers," in *Proceedings of the Third International Conference on Power Quality* (*PQA '93*), San Diego, November 1993.
4. J. Lamoree, D. Mueller, P. Vinett, and W. Jones, "Voltage Sag Analysis Case Studies," in *Proceedings of the 1993 IEEE I&CPS Conference,* St. Petersburg, Fla.
5. M. F. McGranaghan, D. R. Mueller, and M. J. Samotyj, "Voltage Sags in Industrial Systems," *IEEE Transactions on Industry Applications,* vol. 29, no. 2, March/April 1993.
6. Le Tang, J. Lamoree, M. McGranaghan, and H. Mehta, "Distribution System Voltage Sags: Interaction with Motor and Drive Loads," in *Proceedings of the IEEE Transmission and Distribution Conference,* Chicago, April 10–15, 1994, pp. 1–6.
7. EPRI RP 3098-1, *An Assessment of Distribution Power Quality,* Electric Power Research Institute, Palo Alto, Calif.
8. J. Burke, *Power Distribution Engineering: Fundamentals and Applications,* Marcel Dekker, New York, 1994.
9. C. M . Warren, "The Effect of Reducing Momentary Outages on Distribution Reliability Indices," *IEEE Transactions on Power Delivery,* July 1993, pp. 1610–1617.
10. R. C. Dugan, L. A. Ray, D. D. Sabin, et al., "Impact of Fast Tripping of Utility Breakers on Industrial Load Interruptions," in *Conference Record of the 1994 IEEE/IAS Annual Meeting,* Denver, October 1994, vol. III, pp. 2326–2333.
11. T. Roughan and P. Freeman, "Power Quality and the Electric Utility: Reducing the Impact of Feeder Faults on Customers," in *Proceedings of the Second International Conference on Power Quality: End-use Applications and Perspectives* (*PQA '92*), Atlanta, September 28–30, 1992.
12. J. Lamoree, Le Tang, C. De Winkel, and P. Vinett, "Description of a Micro-SMES System for Protection of Critical Customer Facilities," *IEEE Transactions on Power Delivery,* April 1994, pp. 984–991.
13. R. A. Stansberry, "Protecting Distribution Circuits: Overhead Shield Wire Versus Lightning Surge Arresters," *Transmission & Distribution,* April 1991, pp. 56*ff.*

Chapter

4

Transient Overvoltages

4.1 Sources of Transient Overvoltages

There are two main sources of transient overvoltages on utility systems: capacitor switching and lightning. These are also sources of transient overvoltages within end-user facilities as well as a myriad of other switching phenomena. Some power electronic devices generate significant transients when they switch. As described in Chap. 2, transient overvoltages can be high frequency (load switching and lightning), medium frequency (capacitor energizing), or low frequency.

4.1.1 Capacitor switching

Capacitor switching is one of the most common switching events on utility systems. Capacitors are used to provide reactive power (vars) to correct the power factor, which reduces losses, and to support the voltage on the system. They are a very economical and generally trouble-free means of accomplishing this. Alternatives such as rotating machines and electronic var compensators are much more costly or have high maintenance costs. Thus, capacitors are common on power systems and will continue to be.

One drawback to capacitors is that they can interact with the inductance of the power system to yield oscillatory transients when switched. Some capacitors are energized all the time (a fixed bank) while others are switched according to load levels. Various control means are used to determine when they are switched, including time, temperature, voltage, current, and reactive power. It is common for controls to combine two or more of these functions, such as temperature with voltage override.

One of the common symptoms of power quality problems related to utility capacitor-switching overvoltages is that the problems appear at nearly the same time each day. On distribution feeders with industrial loads, capacitors are frequently switched by time clock in anticipation of an increase in load with the beginning of the working day. Common problems are ASD trips and malfunctions of other electronically controlled load equipment without a noticeable blinking of the lights or impact on other, more conventional loads.

Figure 4.1 shows the one-line diagram of a typical utility feeder capacitor-switching situation. When the switch is closed, a transient similar to the one in Fig. 4.2 may be observed upline from the capacitor. As this figure shows, the capacitor switch contacts closed near the system voltage peak, which is common for many switches. The voltage across the capacitor at this instant is zero.

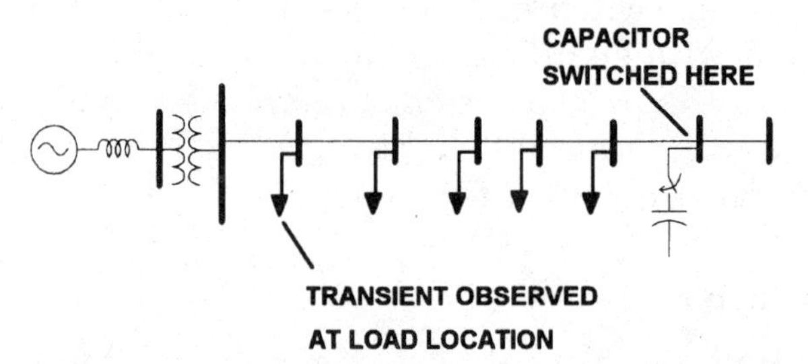

Figure 4.1 One-line diagram of capacitor-switching operation corresponding to the waveform in Fig. 4.2.

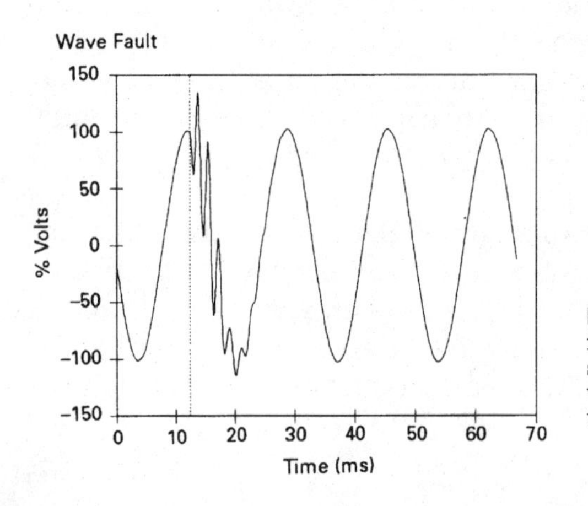

Figure 4.2 Typical electric utility capacitor-switching transient reaching 134 percent voltage, observed upline from the capacitor.

Since it cannot change instantaneously, the system voltage at the capacitor location is briefly pulled down to zero and rises as the capacitor begins to charge toward the system voltage. As is typical of capacitors in inductive power systems, the capacitor voltage overshoots and rings at the natural frequency of the system. If the observation point is upline as indicated, the initial change in voltage will not go completely to zero because of the impedance between the observation point and the switched capacitor.

The overshoot will generate a transient between 1.0 and 2.0 per unit depending on system damping. In this case the transient is about 1.34 per unit. Utility capacitor switching transients are generally in the 1.3 to 1.4 per unit range, but have also been observed near the theoretical maximum.

The transient shown in the oscillogram propagates into the local power system and generally passes through distribution transformers into customer load facilities by an amount nearly equal to the turns ratio of the transformer. If there are capacitors on the secondary system, the voltage may actually be magnified if the natural frequencies of the systems are properly aligned (see next section). While transients up to 2.0 per unit are not generally damaging to the system insulation, the occurrence of such transients can often cause misoperation of electronic power conversion devices. Controllers may interpret the high voltage as a sign that there is an impending dangerous situation and subsequently dump the load to be safe. The transient may also interfere with the gating of thyristors.

Switching of grounded-wye banks may also result in unusual transient voltages in the local ground system due to the current surge that accompanies the energization. Figure 4.3 shows the phase current observed for the capacitor-switching incident described above. The transient current flowing in the feeder peaks at nearly four times the load current.

4.1.2 Magnification of capacitor-switching transients

A potential side effect of adding power factor correction capacitors at the customer location is that they may increase the impact of utility capacitor-switching transients on end-use equipment. There is always a brief voltage transient when capacitor banks are switched. It is generally 1.3 to 1.4 pu of nominal voltage, as demonstrated above, and generally is no more than 2.0 pu. Magnification of the transient overvoltage at the end user bus is pos-

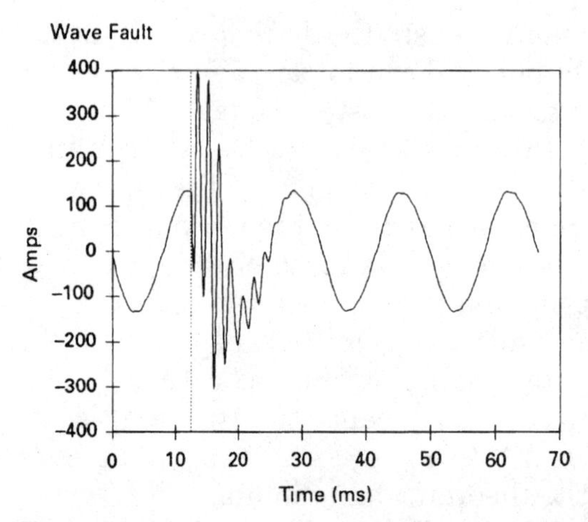

Figure 4.3 Feeder current associated with capacitor-switching event.

sible for certain low-voltage capacitor and step-down transformer sizes. The circuit of concern for this phenomenon is illustrated in Fig. 4.4. Transient overvoltages on the end-user side may reach as high as 3.0 to 4.0 pu on the low-voltage bus under these conditions, with potentially damaging consequences for all types of customer equipment.

Magnification of utility capacitor-switching transients at the end-user location occurs over a wide range of transformer and capacitor sizes. Resizing the customer power factor correction capacitors or step-down transformer is therefore usually not a practical solution. Controlling the transient overvoltage at the utility capacitor is sometimes possible using synchronous closing breakers or switches with preinsertion resistors. These are discussed in more detail in Sec. 4.4.

At the customer location, high-energy surge arresters can be applied to limit the transient voltage magnitude at the customer bus. Energy levels associated with the magnified transient are typically in the range of 1 kJ. Figure 4.5 shows the expected arrester energy for a range of low-voltage capacitor sizes. Newer high-energy metal oxide varistor (MOV) arresters for low-voltage applications can withstand 2 to 4 kJ.

It is important to note that the arresters can only limit the transient to the arrester protective level. This is typically 1.8 times the normal peak voltage (1.8 pu). This level may not be sufficient to protect sensitive electronic equipment that might only have a withstand capability of 1.75 pu [1200-V peak in-

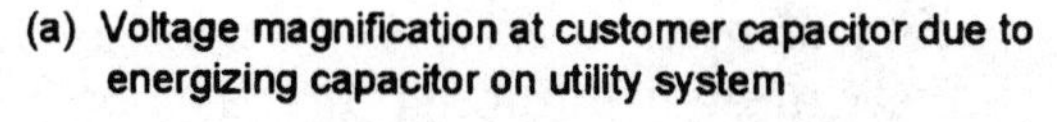

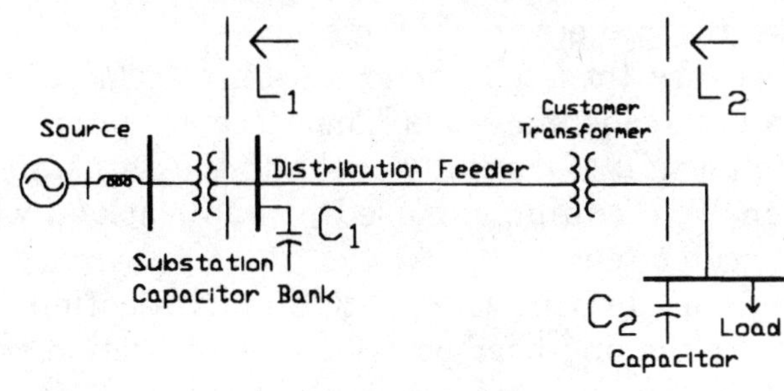

(b) Equivalent circuit

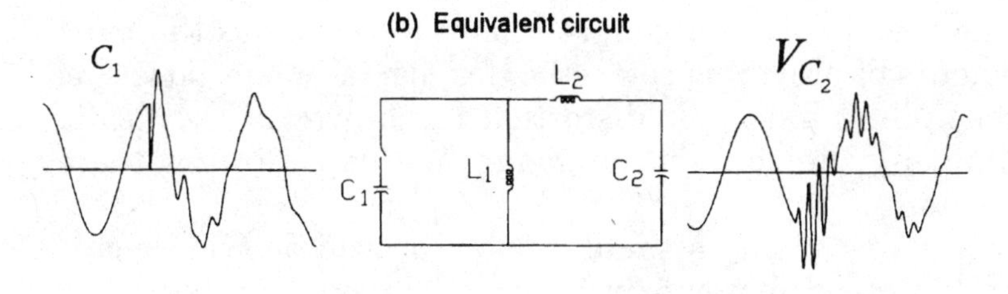

Switching frequency $f_1 = \frac{1}{2\pi\sqrt{L_1C_1}}$

Natural frequency of customer resonant circuit $f_2 = \frac{1}{2\pi\sqrt{L_2C_2}}$

Voltage magnifaction $\Leftrightarrow f_1 \approx f_2$

Figure 4.4 Voltage magnification of capacitor bank switching.

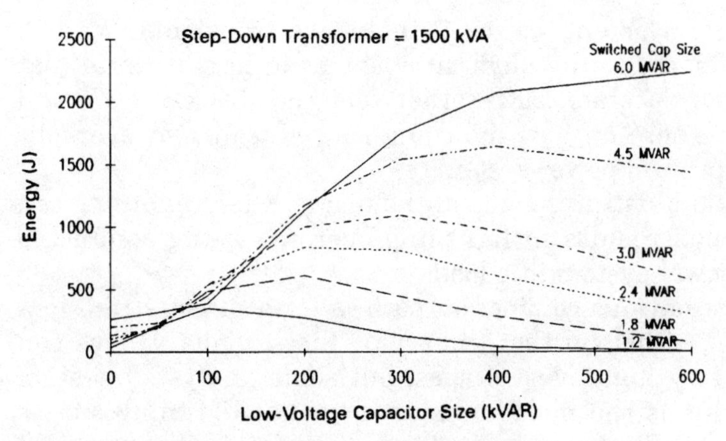

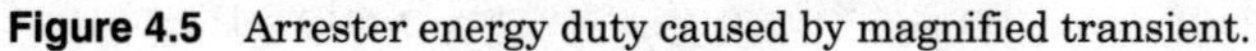
Figure 4.5 Arrester energy duty caused by magnified transient.

verse voltage (PIV) rating of many silicon-controlled rectifiers (SCRs) used in the industrial environment]. It may not be possible to improve the protective characteristics of the arresters substantially because these characteristics are limited by the physics of the metal oxide materials. Therefore, for proper coordination, it is important to carefully evaluate the withstand capabilities of sensitive equipment used in applications where these transients can occur.

Another means of limiting the voltage magnification transient is to convert the end-user power factor correction banks to harmonic filters. An inductance in series with the power factor correction bank will decrease the transient voltage at the customer bus to acceptable levels. This solution has the multiple benefits of providing correction for displacement power factor, controlling harmonic distortion levels within the facility, and limiting the concern for magnified capacitor-switching transients.

In many cases, only a small number of load devices, such as adjustable-speed motor drives, are adversely affected by the transient. It is frequently more economical to place line reactors in series with the drives to block the high-frequency magnification transient. A 3 percent reactor is generally effective. While offering only a small impedance to power frequency current, it offers a considerably larger impedance to the transient. Many types of drives have this protection inherently, either through an isolation transformer or a dc bus reactance.

4.1.3 Lightning

Lightning is a potent source of impulsive transients. We will not devote space to the physical phenomena here because that topic is well documented in other reference books.[1–3] We will concentrate here on how lightning causes transient overvoltages to appear on power systems.

Figure 4.6 illustrates some of the places where lightning can strike, which results in lightning currents being conducted from the power system into loads.

The most obvious conduction path is for a direct strike to a phase wire, either on the primary or the secondary. This can generate very high overvoltages, but some analysts question whether this is the most common way that lightning surges enter load facilities and cause damage. Very similar transient

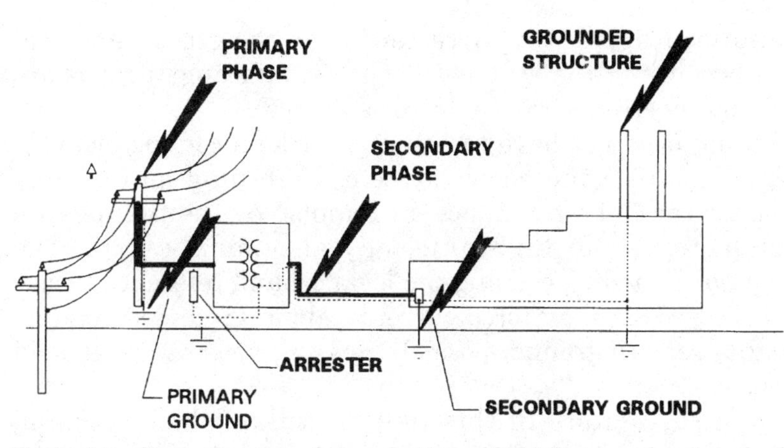

Figure 4.6 Stroke locations for conduction of lightning impulses into load facilities.

overvoltages can be generated by lightning currents flowing along ground conductor paths. Note that there can be numerous paths for lightning currents to enter the grounding system. Common ones, indicated by the dotted lines in Fig. 4.6, include the primary ground, secondary ground, and the structure of the load facilities. Note also that strokes to the primary phase are conducted to the ground circuits through the arresters on the service transformer. Thus, many more lightning impulses may be observed at loads than one might think.

Keep in mind that grounds are never perfect conductors, especially for impulses. While most of the surge current may eventually be dissipated into the ground connection closest to the stroke, there will be substantial surge currents flowing in other connected ground conductors in the first few microseconds of the strike.

A direct strike to phase generally causes line flashover near the strike point. Not only does this generate an impulsive transient, but it also causes a fault with the accompanying voltage sags and interruptions. The lightning surge can be conducted several kilometers along utility lines and may cause multiple flashovers at pole and tower structures as it passes. The interception of the impulse from the phase wire is fairly straightforward by properly installed surge arresters. If the line flashes over at the source of the stroke, the tail of the impulse is generally truncated. Depending on the effectiveness of the grounds along the surge current path, some of the current may find its way into load apparatus. Arresters near the stroke may not survive because of the severe duty (most lightning strokes are actually many strokes in rapid-fire sequence).

Lightning does not have to actually strike a conductor to inject impulses into the power system. Lightning may simply strike near the line and induce an impulse by the the collapse of the electric field. Lightning may also simply strike the ground near a facility, causing the local ground reference to rise considerably. This may force currents along grounded conductors into a remote ground, possibly passing near sensitive load apparatus.

Many investigators in this field postulate that lightning surges enter loads from the utility system through the interwinding capacitance of the service transformer as shown in Fig. 4.7. The concept is that the lightning impulse is so fast that the inductance of the transformer blocks the first part of the wave from passing through by the turns ratio. Instead, the interwinding capacitance offers a ready path for the high-frequency surge. This can permit the existence of a voltage on the secondary terminals that is much higher than what the turns ratio would suggest.

The degree to which capacitive coupling occurs is greatly dependent on the design of the transformer. Not all transformers have a straightforward high-to-low capacitance because of the way the windings are interlaced. The winding-to-ground capacitance may be greater than the winding-to-winding capacitance and more of the surge may actually be coupled to ground than to the secondary winding. In any case, the resulting transient is a very short, single impulse, or train of impulses, because the interwinding capacitance charges quickly. Arresters on the secondary should have no difficulty with the energy in such a surge, but the rates of rise can be high. Thus, lead length be-

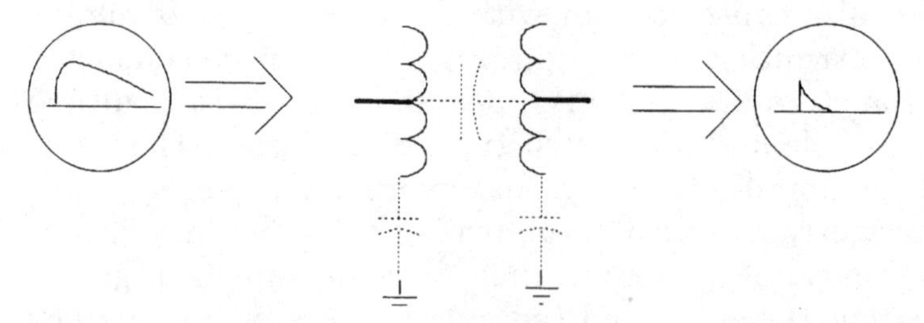

Figure 4.7 Coupling of impulses through the interwinding capacitance of transformers.

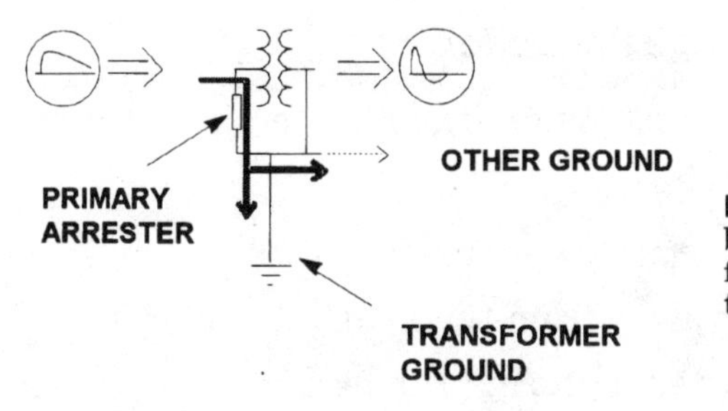

Figure 4.8 Lightning impulse bypassing the service transformer through ground connections.

comes very important to the success of an arrester in keeping this impulse out of load equipment.

Many times, a longer impulse, that is sometimes oscillatory, is observed on the secondary when there is a strike to the utility primary distribution system. This is likely not due to capacitive coupling through the service transformer, but rather to conduction around the transformer through the grounding systems as shown in Fig. 4.8. This is a particular problem if the load system offers a better ground and much of the surge current flows through conductors in the load facility on its way to ground.

The chief power quality problems with lightning stroke currents entering the ground system are that they:

1. Raise the potential of the local ground above other grounds in the vicinity by several kilovolts. Sensitive electronic equipment that is connected between two ground references, such as a computer connected to the telephone system through a modem, can fail when subjected to the lightning surge voltages.
2. Induce high voltages in phase conductors as they pass through cables on the way to a better ground.

The problems are related to the so-called low-side surge problem that is described later in this chapter.

Ideas about lightning are changing with recent research.[8] Lightning causes more flashovers of utility lines than previously thought. Evidence is also mounting that stroke current wavefronts are faster than previously thought and that multiple strikes appear to be the norm rather than the exception. Durations of some strokes may also be longer than reported by earlier researchers. These findings may help explain failures of lightning arresters that were thought to have adequate capacity to handle large lightning strokes.

4.2 Principles of Overvoltage Protection

The fundamental principles of overvoltage protection of load equipment are:

1. Limit the voltage across sensitive insulation
2. Divert the surge current away from the load
3. Block the surge current from entering the load
4. Bond ground references together at the equipment
5. Reduce, or prevent, surge current from flowing between grounds
6. Create a low-pass filter using limiting and blocking principles

These principles are illustrated in Fig. 4.9.

The main function of surge arresters and transient voltage surge suppressors (TVSS) is to limit the voltage that can appear between two points in the circuit. This is an important concept to understand. One of the common misconceptions about varistors, and similar devices, is that they somehow are able to absorb the surge or divert it to ground independently of the rest of the system. That may be a beneficial side effect of the arrester application if there is a suitable path for the surge current to flow into, but the foremost concern in arrester application is to place them directly across the sensitive insulation that is to be protected so that the voltage seen by the insulation is limited to a safe value.

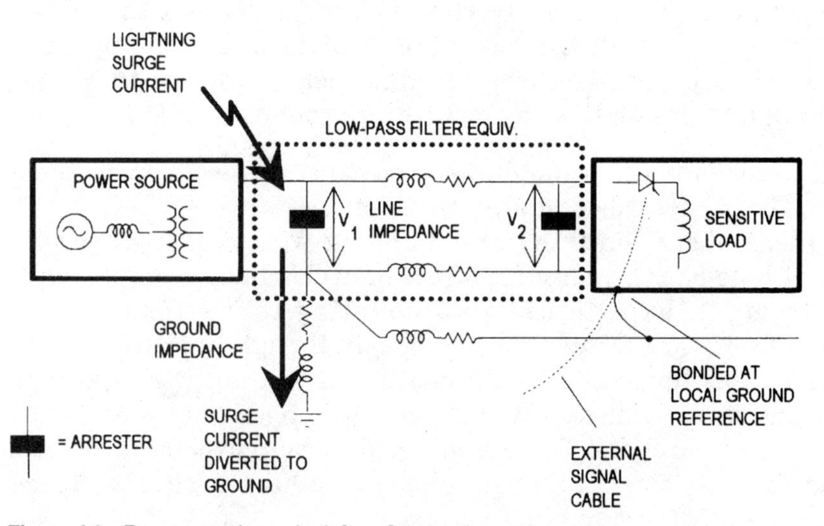

Figure 4.9 Demonstrating principles of protection.

Surge currents must obey Kirchoff's laws just like power current: they must flow in a complete circuit, and they cause a voltage drop in every conductor through which they flow.

One of the points to which arresters, or surge suppressors, are connected is frequently the local ground reference, but it need not be. Keep in mind that the local ground may not be at zero voltage during transient surges.

Surge suppression devices should be located as closely as possible to the critical insulation with a minimum of lead length on all terminals. While it is common to find arresters at panels and subpanels, arresters applied at the point where the power line enters the load equipment are generally the most effective in protecting that particular load. In some cases, the best location is actually inside the load device. For example, many electronic controls made for service in the power system environment have protectors on every line that leaves the cabinet.

In Fig. 4.9, the first arrester is connected from the line to the neutral-ground bond at the service entrance. It limits the line voltage, V_1, from rising too high relative to the neutral and ground voltage at the panel. When it performs its voltage-limiting action, it provides a low-impedance path for the surge current to travel onto the ground lead. Note that the ground lead and the ground connection itself have significant impedance. Therefore, the whole power system is raised in potential with respect to remote ground by the voltage drop across the ground impedance. For common values of surge currents and ground impedances, this can be several kilovolts.

One hopes in this situation that most of the surge energy will be discharged through the first arrester directly into ground. In that sense, the arrester becomes a surge "diverter." This is the another important function related to surge arrester application. In fact, some prefer to call a surge arrester a surge diverter because its voltage limiting action offers a low-impedance path around the load being protected. However, it can only be a diverter if there is a suitable path for the current to be diverted into. That is not always easy to achieve and the surge current is sometimes diverted toward another critical load where it is not wanted.

In this figure, there is another possible path for the surge current to flow into: the signal cable indicated by the dotted line and bonded to the safety ground. If this is connected to another device that is referenced to ground elsewhere, there will be some amount of surge current flowing down the safety ground conductor. Damaging voltages can be impressed across the load as a result. The first arrester at the service entrance is electrically too remote to provide adequate load protection. Therefore, a second arrester is applied at the load—again, directly across the insulation to be protected. It is shown connected line to neutral so that it only protects against normal mode transients. This illustrates the principles without complicating the diagram, but should be considered as the *minimum* one would apply to protect the load. Frequently, surge suppressors will have suppression on all lines to ground, all lines to neutral, and neutral to ground.

While lightning surge currents are seeking a remote ground reference, many transient overvoltages generated by switching will be normal mode and will not seek ground. In cases where surge currents are diverted into other load circuits, arresters must be applied at each load along the path to ensure protection.

Note that the signal cable is bonded to the local ground reference at the load before entering the cabinet. It might seem that this creates an unwanted ground loop. However, it is essential to achieving protection of the load and the low-voltage signal circuits. Otherwise, the power components can rise in potential with respect to the signal circuit reference by several kilovolts. Many loads have multiple power and signal cables connected to them. Also, a load may be in an environment where it is close to another load and operators or sensitive equipment are routinely in contact with both loads. This raises the possibility that a lightning strike may raise the potential of one "ground" much higher than the others. This can cause the flashover of insulation that is between the two ground references or physical harm to operators. Thus, all ground reference conductors (safety grounds, cable shields, cabinets, etc.) should be bonded together at the load equipment. The principle is not to prevent the local ground reference from rising with the surge; with lightning, that is impossible. Rather, the principle is to tie the references together so that all power and signal cable references in the vicinity rise together.

This phenomenon is a common reason for failure of electronic devices. The situation occurs in television receivers connected to cable, computers connected to modems, computers with widespread peripherals powered from various sources, and in manufacturing facilities with networked machines.

Since a few feet of conductor make a significant difference at lightning surge frequencies, it is sometimes necessary to create a special low-inductance ground reference plane for sensitive electronic equipment such as mainframe computers that occupy large spaces.[4]

Efforts to block the surge current are most effective for high-frequency surge currents such as those originating with lightning strokes and capacitor-switching events. Since power frequency currents must pass through the surge suppressor with minimal additional impedance, it is difficult and expensive to build filters that are capable of discriminating between low-frequency surges and power frequency currents.

Blocking can be done relatively easily for high-frequency transients by placing an inductor, or choke, in series with the load. The high surge voltage will be dropped across the inductor. One must be careful to consider how high the voltage might get across the inductor because it is possible to fail its insulation as well as that of the loads. However, a line choke alone is frequently an effective means to block such high-frequency transients as line-notching transients from adjustable-speed drives.

The blocking function is frequently combined with the voltage-limiting function to form a "low-pass filter" in which there is a shunt-limiting device on either side of the series choke. Figure 4.9 illustrates how such a circuit naturally occurs when there are arresters on both ends of the line feeding the load. The line provides the blocking function in proportion to its length. Such a circuit has very beneficial overvoltage protection characteristics. The inductance forces the bulk of fast-rising surges into the first arrester. The second arrester then simply has to clean up what little surge energy gets through. Such circuits are also commonly built into outlet strips for computer protection.

A large percentage of surge protection problems occur because the surge current travels between two, or more, separate connections to ground. This is a particular problem with lightning protection because lightning currents are seeking ground and basically divide according to the ratios of the impedances of the ground paths. The surge current does not even have to enter the power, or phase, conductors to cause problems. There will be a significant voltage drop along the ground conductors that will frequently appear across critical insulation. The grounds involved may be entirely within the load facility or some of the grounds may be on the utility system.

Ideally, there would be only one ground path for lightning within a facility, but many facilities have multiple paths. For example, there may be a driven ground at the service entrance or substation transformer and a second ground at a water well that actually creates a better ground. Thus, when lightning strikes, the bulk of the surge current tends to flow toward the well. This can impress an excessively high voltage across the pump insulation. Such a situation can occur even if the electrical system is not intentionally bonded to a second ground. When lightning strikes, the potentials can become so great that the power system insulation will flash over somewhere.

The amount of current flowing between the grounds may be reduced by improving all the intentional grounds at the service entrance and nearby on the utility system. This will normally reduce, but not eliminate entirely, the incidence of equipment failure within the facility due to lightning. However, some structures also have significant lightning exposure and the damaging surge currents can flow back into the utility grounds. It doesn't matter in which direction the currents flow; they cause the same problems wherever they flow. Again, the same principle applies: Improve the grounds for the structure to minimize the amount of current that might seek another path to ground.

When it is impractical to keep the currents from flowing between two grounds, *both* ends of any power or signal cables running between the two grounds must be protected with voltage-limiting devices to ensure adequate protection. Circuits in this condition are found commonly in utility and end-user systems alike where a control cabinet is quite some distance from the switch, or other device, being controlled.

4.3 Devices for Overvoltage Protection

Surge arresters and TVSS are devices that protect equipment from transient overvoltages by limiting the maximum voltage. The terms *arrester* and *TVSS* are sometimes used interchangeably. However, arrester is most frequently used to describe the device in the service entrance while TVSS is generally used to describe a similar device used at the load equipment. A TVSS sometimes has more surge-limiting elements than an arrester, which most commonly consists of MOV blocks. A device called an arrester may have more energy-handling capability. However, the distinction between the two is blurred by common language usage.

The elements that make up these devices can be classified by two different modes of operation, *crowbar* and *clamping.*

Crowbar devices are normally open devices which conduct during overvoltage transients. Once the device conducts, the line voltage drops to nearly zero due to the "short circuit" imposed across the line. These devices are usually manufactured with a gap filled with air or a special gas. The gap arcs over when a sufficiently high overvoltage transient appears. Once the gap arcs over, power frequency current, or "follow current," will continue to flow in the gap until the next current zero. Thus, these devices have the disadvantage that the power frequency voltage drops to zero or to a very low value for at least one-half cycle. This will cause some loads to drop off line unnecessarily.

Clamping devices for ac circuits are commonly nonlinear resistors (varistors) which conduct very low amounts of current until an overvoltage occurs. Then they start to conduct heavily and their impedance drops rapidly with increasing voltage. These devices effectively conduct increasing amounts of current (and energy) to limit the voltage rise of a surge. They have an advantage over gap-type devices in that the voltage is not reduced below the conduction level when they begin to conduct the surge current. Zener diodes are also used in this application. Example characteristics of MOV arresters for load systems are shown in Figs. 4.10 and 4.11.

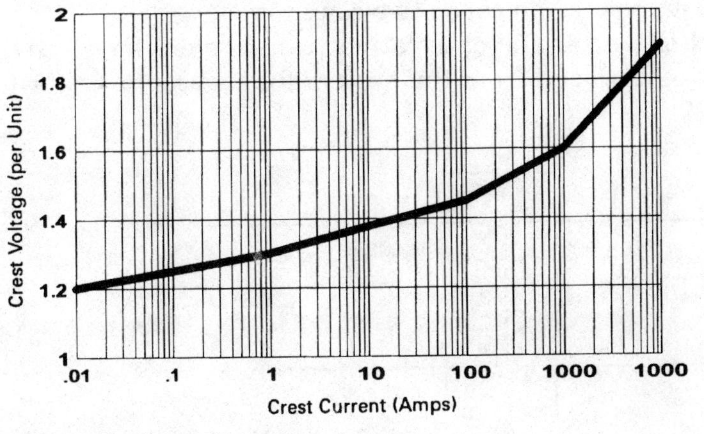

Figure 4.10 Crest voltage vs. crest amps.

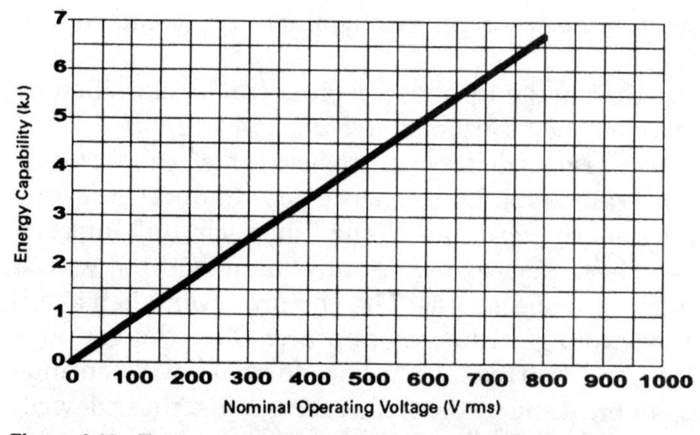

Figure 4.11 Energy capability vs. operating voltage.

MOV arresters have two important ratings. The first is maximum continuous operating voltage (MCOV), which must be higher than the line voltage and is often at least 125 percent of the system nominal voltage. The second rating is the energy dissipation rating (in joules). MOVs are available in a wide range of energy ratings. Figure 4.11 shows the typical energy-handling capability vs. operating voltages.

Isolation transformers (Fig. 4.12) are used to attenuate high-frequency noise and transients as they attempt to pass from one side to the other. However, some common mode and normal mode noise can still reach the load. An electrostatic shield, as shown in Fig. 4.13, is effective in eliminating common mode noise. However, some normal mode noise can still reach the load due to magnetic and capacitive coupling.

The chief characteristic of isolation transformers for electrically isolating the load from the system for transients is their

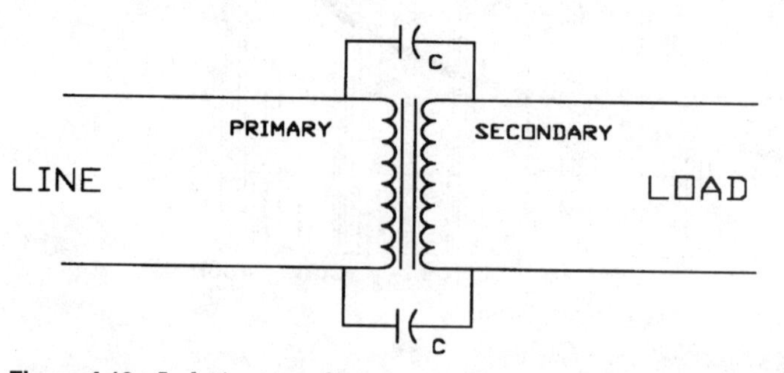

Figure 4.12 Isolation transformer.

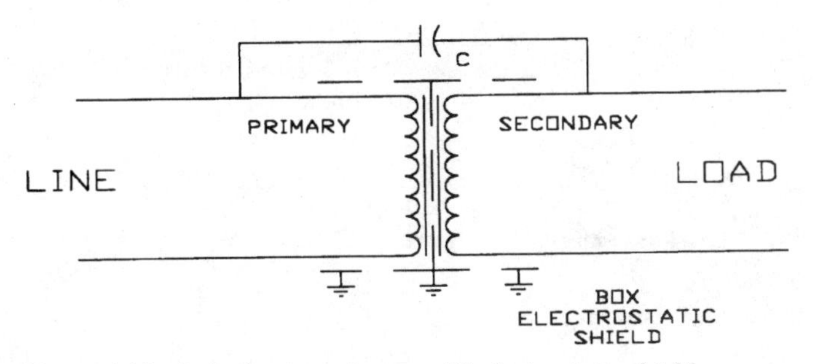

Figure 4.13 Isolation transformer with electrostatic shield.

leakage inductance. Therefore, high-frequency noise and transients are kept from reaching the load, and any load-generated noise and transients are kept from reaching the rest of the power system. Voltage notching due to power electronic switching is one example of a problem that can be limited to the load side by an isolation transformer. Capacitor switching and lightning transients coming from the utility system can be attenuated, thereby preventing nuisance tripping of adjustable-speed drives and other equipment.

An additional use of isolation transformers is that they allow the user to define a new ground reference, or *separately derived system*. This new neutral-to-ground bond limits neutral-to-ground voltages at sensitive equipment.

Low-pass filters use the pi-circuit principle illustrated in Fig. 4.9 to achieve even better protection for high-frequency transients. For general usage in electrical circuits, low-pass filters are composed of series inductors and parallel capacitors. This LC combination provides a low-impedance path to ground for selected resonant frequencies. In surge protection usage, voltage-clamping devices are added in parallel to the capacitors. In some designs, there are no capacitors.

Figure 4.14 shows a common hybrid protector that combines two surge suppressors and a low-pass filter to provide maximum protection. It uses a gap-type protector on the front end to handle high-energy transients. The low-pass filter limits transfer of high-frequency transients. The inductor helps block high-frequency transients and forces them into the first suppressor. The capacitor limits the rate-of-rise while the nonlinear resistor (MOV) clamps the voltage magnitude at the protected equipment.

Other variations on this design employ MOVs on both sides of the filters and may have capacitors on the front end as well.

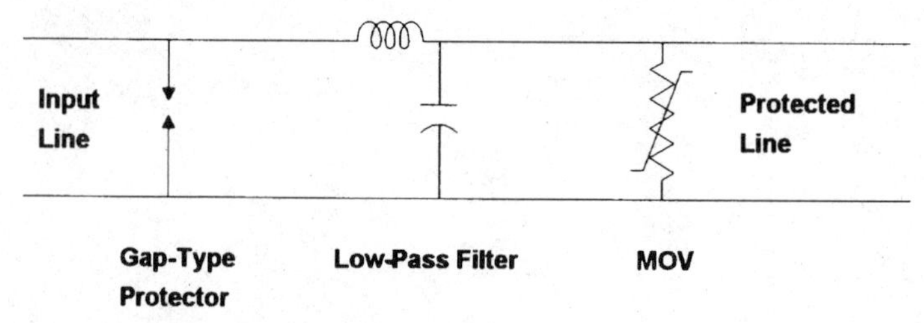

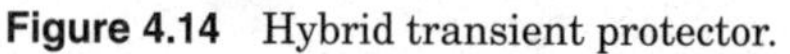

Figure 4.14 Hybrid transient protector.

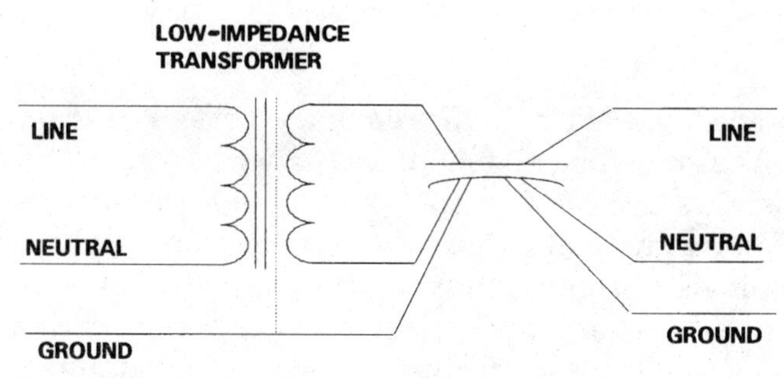

Figure 4.15 Low-impedance power conditioner.

Low-impedance power conditioners (*LIPCs*) are used primarily to interface with the switch-mode power supplies found in electronic equipment. LIPCs differ from isolation transformers in that these conditioners have a much lower impedance, and have a filter as part of their design (Fig. 4.15). The filter is on the output side and protects against high-frequency source-side common mode and normal mode disturbances (i.e., noise and impulses). Note the new neutral-to-ground connection that can be made on the load side because of the existence of an isolation transformer. However, low- to medium-frequency transients (capacitor switching) can cause problems for LIPCs: The transient can be magnified by the output filter capacitor.

4.4 Utility Capacitor-Switching Transients

This section describes how utilities can deal with problems related to capacitor-switching transients.

4.4.1 Switching times

Capacitor-switching transients are very common and usually not damaging. However, timing of switching may be unfortunate for some sensitive industrial loads. For example, the utility might notice that the load picks up at a certain time each day and decides to switch the capacitors coincident with that load increase. There have been several cases where this coincides with the beginning of a work shift and causes several adjustable-speed drives to shut down shortly after the process has been started. One simple and inexpensive solution is to meet with affected end users and determine if there is a time that might be acceptable. For example, it may be possible to switch the capacitor on a few minutes before the beginning of the shift and before the load actually picks up. Switching the capacitor on may not be needed then, but probably won't hurt anything. If this solution cannot be worked out, other, more expensive solutions will have to be found.

4.4.2 Preinsertion resistors

Preinsertion resistors can reduce the capacitor-switching transient considerably. The first peak of the transient is usually the most damaging. The idea is to insert a resistor into the circuit briefly so that the first peak is damped significantly.

Figure 4.16 shows one example of a capacitor switch with preinsertion resistors to reduce transients. The preinsertion is accomplished by the movable contacts sliding past the resistor contacts first before mating with the main contacts. This results in a preinsertion time of approximately 0.25 cycle at 60 Hz. The effectiveness of the resistors is dependent on capacitor size and available short circuit current at the capacitor location. Table 4.1 shows expected maximum transient overvoltages upon energization for various conditions, both with and without the preinsertion resistors. These are the maximum values expected; average values are typically 1.3 to 1.4 per unit without resistors and 1.1 to 1.2 per unit with resistors.

Switches with preinsertion reactors have also been developed for this purpose. The inductor is helpful in limiting the higher-frequency components of the transient. In some designs, the reactors are intentionally lossy so that the energization transients damp out quickly.

4.4.3 Synchronous closing

Another strategy for reducing transients on capacitor-switching is to use a synchronous closing breaker. Synchronous closing

Figure 4.16 Capacitor switch with preinsertion resistors. (*Photograph courtesy of Cooper Power Systems.*)

TABLE 4.1 Peak Transient Overvoltages Due to Capacitor Switching with or without Preinsertion Resistor

Size (kvar)	Available short circuit (kA)	Without resistor (per unit)	With 6.4-Ω resistor (per unit)
900	4	1.95	1.55
900	9	1.97	1.45
900	14	1.98	1.39
1200	4	1.94	1.50
1200	9	1.97	1.40
1200	14	1.98	1.34
1800	4	1.92	1.42
1800	9	1.96	1.33
1800	14	1.97	1.28

Courtesy of Cooper Power Systems.

prevents transients by timing the contact closure so that the system voltage closely matches the capacitor voltage at the instant the contacts mate. This avoids the step change in voltage that normally occurs when capacitors are switched, causing the circuit to oscillate.

Figure 4.17 shows one example of a circuit breaker designed for this purpose. This breaker would normally be applied at the utility subtransmission or transmission system (72- and 145-kV classes). This is a three-phase SF6 breaker that uses a specially designed operating mechanism with three independently controllable drive rods. It is capable of closing within 1 ms of voltage zero. The electronic control samples variables such as ambient temperature, control voltage, stored energy, and the time since the last operation to compensate the algorithms for the timing forecast. The actual performance of the breaker is sampled to adjust the pole timing for future operations to compensate for wear and changes in mechanical characteristics.

Figure 4.18 shows another switch made for this purpose. This is a vacuum switch, applied on distribution capacitor banks in the 15-to-230-kV classes. It consists of three independent poles with separate operating mechanisms. The timing for synchronous closing is determined by anticipating an upcoming voltage zero. Its success is dependent on the consistent operation of the vacuum switch which normally closes within 0.25 ms of voltage zero. The switch reduces capacitor inrush currents by an order of magnitude and voltage transients to about 1.1 per unit.

4.4.4 Capacitor location

For distribution feeder banks, a switched capacitor may be too close to a sensitive customer or at a location where the transient overvoltages tend to be much higher. Often it may be possible to move the capacitor downline or to another branch of the circuit and eliminate the problem. The strategy is either to create some more damping by getting more resistance into the circuit or to get more impedance between the capacitor and the sensitive customer.

The success of this strategy depends on a number of factors. Of course, if the capacitor is placed at a large load to supply reactive power specifically for that load, moving the bank may not be an option. Then techniques for soft switching or switching at noncritical times must be explored.

Figure 4.17 Synchronous closing breaker. (*Photograph courtesy of ABB Power T&D Co., Inc.*)

4.5 Utility Lightning Protection

Many power quality problems stem from lightning. Not only can the high-voltage impulses damage load equipment, but the temporary fault that follows a lightning strike to the line can cause sags and interruptions. Here are some strategies for utilities to use to decrease the impact of lightning.

Figure 4.18 69-kV synchronous closing capacitor switch. (*Photograph courtesy of Joslyn Hi-Voltage Corp.*)

4.5.1 Shielding

One of the strategies open to utilities for lines that are particularly susceptible to lightning strikes is to add shielding to capture most lightning strokes before they strike the phase wires. This can help, but will not necessarily prevent line flashovers because of the possibility of backflashovers.

Shielding of overhead utility lines is common on transmission voltage levels and in substations. It is not common on distribution lines because of the cost associated with placing the grounded neutral wire over the phase wires and the lower benefit due to lower flashover levels of the lines. The grounded neutral wire is typically suspended underneath the phase conductors to facilitate the connection of line apparatus such as transformers and capacitors. Another factor is that taller poles are required for shielding. Thus, shielding adds considerable expense that is not justifiable in the average case.

The application is not quite as simple as adding a wire and grounding it every few poles. When lightning strikes the shield wire, the voltages at the top of the pole are still extremely high and could cause backflashovers to the line. This will also result in a temporary fault. To minimize this possibility, the path of the ground lead down the pole must be carefully chosen to maintain adequate clearance with the phase

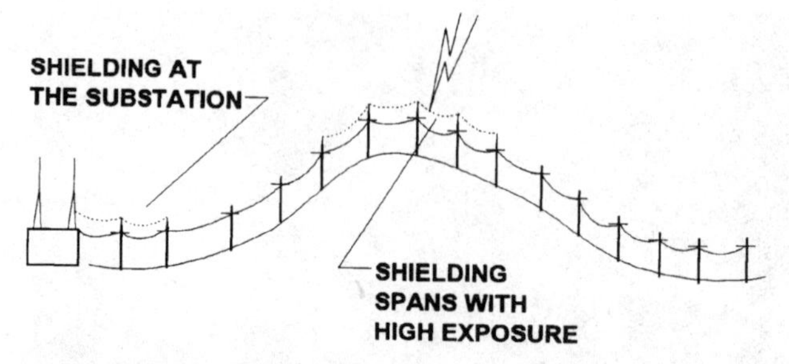

Figure 4.19 Shielding a portion of a distribution feeder to reduce the incidence of temporary lightning-induced faults.

conductors. Also, the grounding resistance plays an important role in the magnitude of the voltage and must be maintained as low as possible.

When it becomes obvious that a particular section of feeder is being struck frequently, it may be justifiable to retrofit that section with a shield wire to reduce the number of transient faults and to maintain a higher level of power quality. Figure 4.19 illustrates this concept. It is not uncommon for a few spans near the substation to be shielded. The substation is generally shielded anyway and this helps prevent high-current faults close to the substation that can damage the substation transformer and breakers. Another section of the feeder may crest a ridge, giving it unusual exposure to lightning. Shielding in that area may be an effective way of reducing lightning-induced faults. Poles in the affected section may have to be extended to accommodate the shield wire. Line arresters may be a more economical and effective option for most applications.

4.5.2 Line arresters

Another strategy for lines that are struck frequently is to apply arresters periodically along the phase wire. Normally, lines flash over first at the pole insulators. Therefore, preventing insulator flashover will reduce the interruption and sag rate significantly. Stansberry[5] argues that this is more economical than shielding.

As shown in Fig. 4.20, the arresters bleed off some of the stroke current as it passes along the line. The amount that an individual arrester bleeds off will depend on the grounding resistance. The idea is to space the arresters sufficiently close to prevent the voltage at unprotected poles in the middle from exceeding the basic impulse level (BIL) of the line insulators.

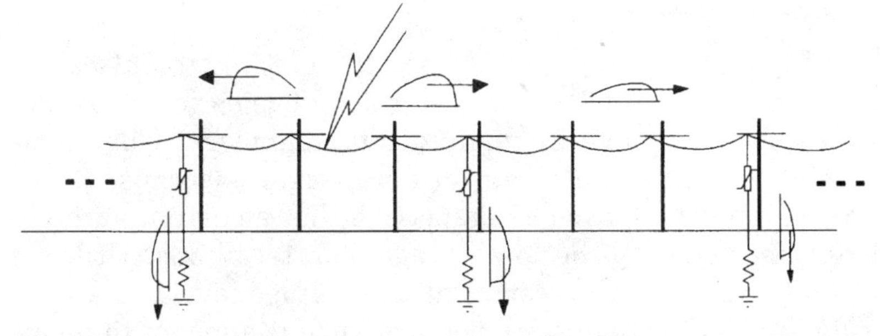

Figure 4.20 Arresters bleed off some of the stroke current as it passes along the line.

This usually requires placing an arrester at every second or third pole. In the case of a feeder supplying a highly critical load, or a feeder with high ground resistance, it may be necessary to place arresters at every pole. A transients study of different configurations will show what is required.

There are already sufficient arresters on many lines in densely populated areas to approximate the desired condition. Since arresters are applied on nearly all distribution transformers in most parts of North America, a transformer on every second or third pole often suffices to provide adequate line protection. Figure 4.21 shows a typical utility arrester used for this application.

Figure 4.21 Typical utility distribution arrester. (*Photograph courtesy of Cooper Power Systems.*)

4.5.3 Low-side surges

Some utility and customer problems with lightning impulses are closely related. One of the most significant ones is called the "low-side surge" problem by many utility engineers.[6] The name was coined by distribution transformer designers because it appears from the transformer's perspective that a current surge is suddenly injected into the low-voltage side terminals. Utilities do not apply secondary arresters at low voltage levels as a general rule. From the customer's point of view it appears to be an impulse coming from the utility and is likely to be termed a "secondary surge."

Both problems are actually different side effects of the same surge phenomenon: The lightning current flowing from either the utility side or the customer side along the service cable neutral. Figure 4.22 shows one possible scenario. Lightning strikes the primary line and the current is discharged through the primary arrester to the pole ground lead. This lead is also connected to the X2 bushing of the transformer at the top of the pole. Thus, some of the current will flow toward the load ground. The amount of current into the load ground is primarily dependent on the size of the pole ground resistance relative to the load ground. Inductive elements may play a significant role in the current division for the front of the surge, but the

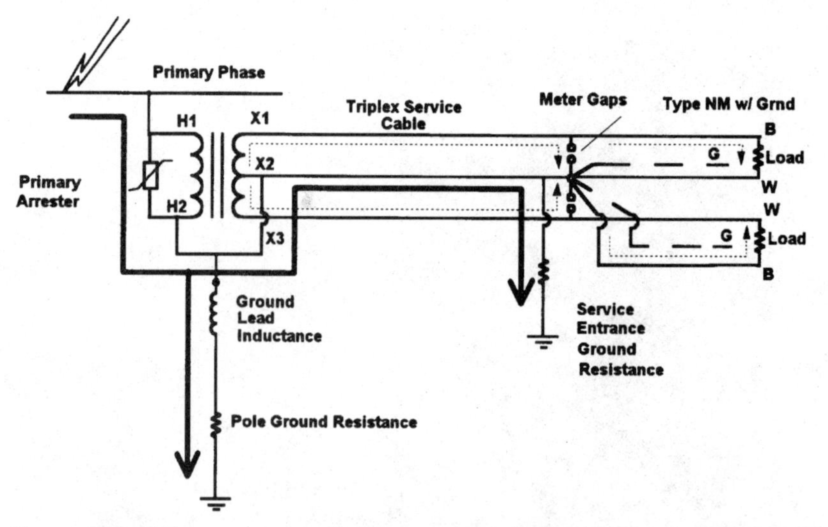

Figure 4.22 Primary arrester discharge current divides between pole and load ground.

ground resistances basically dictate the division of the bulk of the stroke current.

The current that flows through the secondary cables causes a voltage drop in the neutral conductor that is only partially compensated by mutual inductive effects with the phase conductors. Thus, there is a net voltage forcing current through the transformer secondary windings and into the load as shown by the dashed lines in the figure. If there is a complete path, substantial surge current will flow. As it flows through the transformer secondary, a surge voltage is induced in the primary, sometimes causing a layer-to-layer failure near the grounded end. If there is not a complete path, the voltage builds up across the load and may flash over somewhere on the secondary. It is common for the meter gaps to flash over, but not always before there is damage on the secondary because the meter gaps are usually 6 to 8 kV, or higher.

The amount of voltage induced in the cable is dependent on the rate-of-rise of the current, which is dependent on other circuit parameters as well as the lightning stroke.

The chief power quality problems this causes are:

1. The impulse entering the load can cause failure or misoperation of load equipment.
2. The utility transformer will fail, causing an extended power outage.
3. The failing transformer may subject the load to sustained steady-state overvoltages because part of the primary winding is shorted, decreasing the transformer ratio. Failure usually occurs in seconds, but has been known to take hours.

The key to this problem is the amount of surge current traveling through the secondary service cable. Keep in mind that the same effect occurs regardless of the direction of the current. All that is required is for the current to get into the ground circuits and for a substantial portion to flow through the cable on its way to another ground. Thus, lightning strikes to either the utility system or the end-user facilities can produce the same symptoms. Transformer protection is more of an issue in residential services, but the secondary transients appear in industrial systems as well.

Protecting the transformer. There are two common ways for the utility to protect the transformer:

1. Use transformers with interlaced secondary windings.
2. Apply surge arresters at the X terminals.

Of course, the former is a design characteristic of the transformer and cannot be changed once the transformer has been made. If the transformer is a noninterlaced design, the only option is to apply arresters to the low-voltage side.

Note that arresters at the load service entrance will not protect the transformer. In fact, they will virtually guarantee that there will be a surge current path and thereby cause additional stress on the transformer.

While interlaced transformers have a lower failure rate in lightning-prone areas than noninterlaced transformers, recent evidence suggests that low-voltage arresters have better success in preventing failures.[7] Figure 4.23 shows an example of a well-protected utility pole-top distribution transformer. The primary arrester is mounted directly on the tank with very short lead lengths. With the evidence mounting that lightning surges have steeper fronts than previously believed, this is an ever increasing requirement for good protection practice.[8] A special fuse in the cutout is required to prevent fuse damage on

Figure 4.23 Example of a distribution transformer well protected against lightning with both primary and secondary arrester protection. (*Photograph courtesy of Cooper Power Systems.*)

lightning current discharge. The transformer protection is completed by a robust secondary arrester. This shows a heavy-duty secondary arrester adapted for external mounting on transformers. Internally mounted arresters are also available. An arrester rating of 40 kA discharge current is recommended. The voltage discharge is not extremely critical in this application, but is typically 3 to 5 kV. Transformer secondaries are generally assumed to have a BIL of 20 to 30 kV. Gap-type arresters also work in this application, but cause voltage sags, which the MOV-type arrester avoids.

Impact on load circuits. Figure 4.24 shows a waveform of the open-circuit voltage measured at an electrical outlet location in a laboratory mock-up of a residential service.[10] For a relatively small stroke to the primary line (2.6 kA), the voltages at the outlet reached nearly 15 kV. In fact, higher-current strokes caused random flashovers of the test circuit, which made measurements difficult. This reported experience is indicative of the capacity of these surges to cause overvoltage problems.

The waveform is a very high-frequency ringing wave riding on the main part of the low-side surge. The ringing is very sensitive to the cable lengths. A small amount of resistive load such as a light bulb would contribute greatly to the damping. The ringing wave differs depending on where the surge was applied, while the base low-side surge wave remains about the

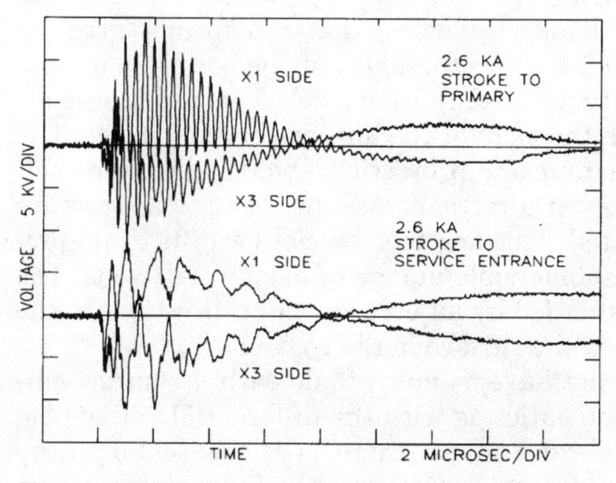

Figure 4.24 Voltage appearing at outlet due to low-side surge phenomena.

same; it is more dependent on the waveform of the current through the service cable.

One interesting aspect of this wave is that the ringing is so fast that it gets by the spark gaps in the meter base even though the voltage is two times the nominal sparkover value. In the tests, the outlets and lamp sockets could also withstand this kind of wave for about 1 μs before they flashed over. Thus, it is possible that some high overvoltages can propagate throughout the system. The waveform in this figure represents the available open-circuit voltage. In actual practice, a flashover would have occurred somewhere in the circuit after a brief time.

MOV arresters are not entirely effective against a ringing wave of this high frequency because of lead length inductance. However, they are very effective for the lower-frequency portion of this transient, which contains the greater energy. Arresters should be applied both in the service entrance and at the outlets serving sensitive loads. Without the service entrance arresters to take most of the energy, arresters at the outlets are subject to failure. This is particularly true of single MOVs connected line-to-neutral. With service entrance arresters in place, failure of outlet protectors and individual appliance protectors should be very rare unless lightning strikes nearer to them than to the service entrance.

Service entrance arresters cannot be relied upon to protect the entire facility. They serve a useful purpose in shunting the bulk of the surge energy, but cannot suppress the voltage sufficiently for remote loads. Likewise, the transformer arrester cannot be considered to take the place of the service entrance arrester although it may be only 15 m away. This arrester is actually in *series* with the load for the low-side current surge. The basic guideline for arrester protection should always be followed: place an arrester directly across the insulation structure that is to be protected. This becomes crucial for difficult-to-protect loads such as submersible pumps in deep water wells. The best protection is afforded by an arrester built directly into the motor rather than on the surface in the controller.

Some cases may not have as much to do with the surge voltage appearing at the outlet as with the differential voltage between two ground references. Such is the case for many television receiver failures. Correct *bonding* of protective grounds is required as well as arrester protection.

The protective level of service entrance arresters for lightning impulses is typically about 2 kV. The lightning impulse current-carrying capability should be similar to the transformer secondary arrester, or approximately 40 kA. One must keep in mind that for low-frequency overvoltages, the arrester with the lowest discharge voltage is apt to take the brunt of the duty. MOV-type arresters clamp the overvoltages without causing additional power quality problems such as interruptions and sags.

4.5.4 Cable protection

One increasingly significant source of extended power outages is UD cable failures. The earliest utility distribution cables installed in the United States are now reaching the end of their useful life. As a cable ages, the insulation becomes progressively weaker and a moderate transient overvoltage causes breakdown and failure.

Many utilities are exploring ways of extending the cable life by arrester protection. Cable replacement is so costly that it is often worthwhile to retrofit the system with arresters even if the gain in life is only a few years. Depending on voltage class, the cable may have been installed with only one arrester on the riser pole or both a riser-pole arrester and an open-point arrester (see Fig. 4.25).

To provide additional protection, utilities may choose from a number of options:

1. Add an open-point arrester, if one does not exist.
2. Add a third arrester on the next-to-last transformer.
3. Add arresters at every transformer.
4. Add special low-discharge voltage arresters.
5. Inject an insulation-restoring fluid into the cable.

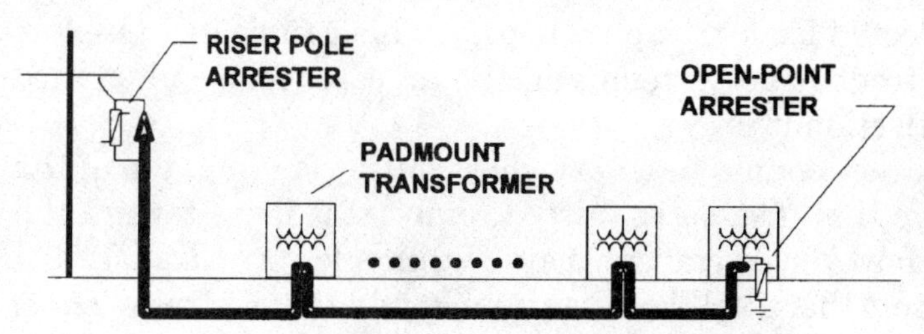

Figure 4.25 Typical underground cable arrester application.

6. Employ a scout arrester scheme on the primary (see Sec. 4.5.5).

The cable life is an exponential function of the number of impulses of a certain magnitude that it receives, according to Hopkinson.[9] The damage to the cable is related by

$$D = NV^{c}$$

where D = constant, representing damage to the cable
N = number of impulses
V = magnitude of impulses
c = empirical constant ranging from 10 to 15

Therefore, anything that will decrease the magnitude of the impulses only slightly has the potential to extend cable life a great deal.

Open-point arrester. Voltage waves double in magnitude when they strike an open point. Thus, the peak voltage appearing on the cable is about twice the discharge voltage of the riser-pole arrester. There is sufficient margin with new cables to get by without open-point arresters at some voltage classes. While open-point arresters are common at 35 kV, they are not used universally at lower voltage classes.

When the number of cable failures associated with storms begins to increase noticeably, the first option should be to add an arrester at the open point if there is not already one present.

Next-to-last transformer. Open-point arresters do not completely eliminate cable failures during lightning storms. With an open-point arrester, the greatest overvoltage stress is generally found at the next-to-last transformer. Figure 4.26 illustrates the phenomenon. Before the open-point arrester begins to conduct, it reflects the incoming wave just like an open circuit. Therefore, there is a wave of approximately half the discharge voltage reflected back to the riser pole. This can be even higher if the wavefront is very steep and the arrester lead inductance aids the reflection briefly.

This situation results in a very short pulse riding on top of the voltage wave that dissipates fairly rapidly as it flows toward the riser pole. However, at transformers within a few hundred feed of the open point there will be a noticeable additional stress. Thus, we often see cable and transformer failures at this location.

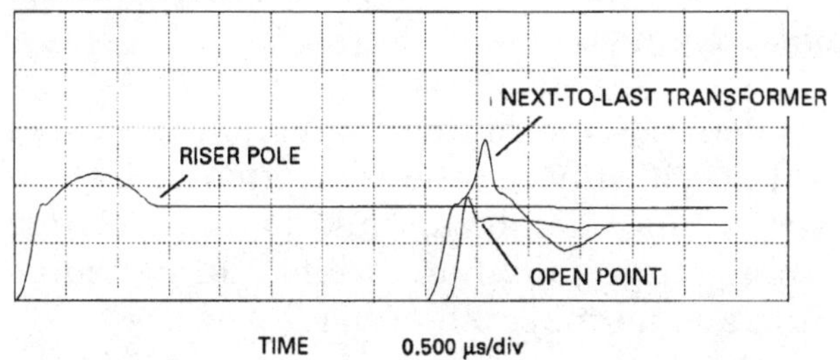

Figure 4.26 Impulse voltages along cable with an open-point arrester showing that the peak can occur at the next-to-last transformer. Simulation with UDSurge™ computer program. (*Courtesy of Cooper Power Systems.*)

The problem is readily solved by an additional arrester at the next-to-last transformer. In fact, this second arrester practically obliterates the impulse, providing effective protection for the rest of the cable system as well. Thus, some consider the most optimal UD cable protection configuration to be three arresters: a riser-pole arrester, an open-point arrester, and an arrester at the transformer next closest to the open point. This choice is nearly as good as having arresters at all transformers and is less costly, particularly in retrofitting.

Under-oil arresters. Transformer manufacturers can supply pad-mounted transformers for UD cable systems with the primary arresters inside the transformer compartment, under oil. If applied consistently, this achieves very good protection of the UD cable system by having arresters distributed along the cable. Of course, this protection comes at an incremental cost that must be evaluated to determine if it is economical for a utility to consider.

Elbow arresters. The introduction of elbow arresters for transformer connections in UD cable systems has opened up protection options not previously economical. Previously, arrester installations on UD cable systems were adaptations of overhead arrester technology and were costly to implement. That is one reason why open-point arresters have not been used universally. The other alternative was under-oil arresters and it is also very costly to change out a padmount transformer just to get an open-point arrester. Now, the arrester is an integral part of the UD system hardware and installation at nearly any point on the system is practical. This is a particularly good option for many retrofit programs.

Lower discharge arresters. Some newer-technology arresters have been developed specifically for the protection of UD cables. The goal was to achieve a substantially lower discharge voltage under lightning surge conditions while still providing the capability to withstand normal system conditions.

One technology that is now being promoted for this is the gapped MOV arrester. In the 1970s, utility suppliers began to replace gapped silicon carbide arresters with gapless MOV arresters. The gaps were the source of many of the problems associated with arrester functionality, and there was a strong desire to get rid of them. Silicon carbide required gaps to prevent currents from flowing at normal operating voltages. MOV technology could withstand normal system voltages while providing similar discharge voltages and more reliable operation. By mixing gaps and MOV technology, a 20 to 30 percent gain may be made in protective margin. The gaps share the voltage during steady-state operation and prevent thermal runaway in the MOV.

Following the logic of the Hopkinson formula described above, applying this kind of arrester in the UD cable system can be expected to yield a substantial gain in life over conventional arrester technology.

Fluid injection. This is a relatively new technology in which a restorative fluid is injected into a run of cable. The fluid fills the voids that have been created in the insulation by aging and gives the cable many more years of life. A vacuum is pulled on the receiving end and pressure is applied at the injection end. If there are no splices to block the flow, the fluid slowly penetrates the cable.

4.5.5 Scout arrester scheme

The idea of using a scout arrester scheme to protect utility UD cable runs goes back many years.[11] However, the idea has only been applied sporadically because of the additional initial expense. The concept is relatively simple: place arresters on either side of the riser-pole arrester to reduce the lightning energy that can enter the cable. Figure 4.27 illustrates the

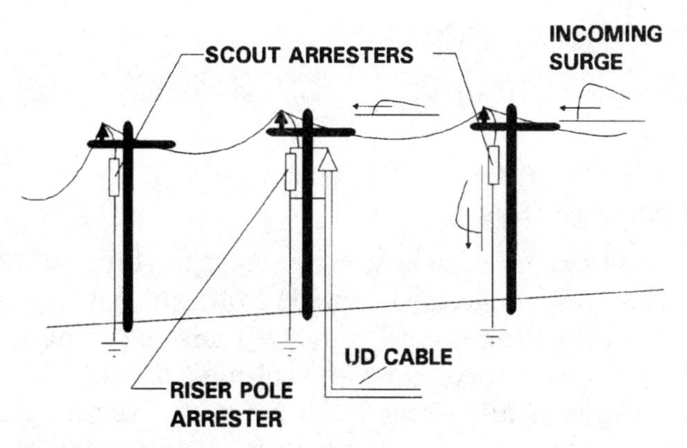

Figure 4.27 Scout arrester scheme.

basic scheme. The incoming lightning surge current from a strike downline first encounters a scout arrester. A large portion of the current is discharged into the ground at that location. A smaller portion proceeds on to the the riser-pole arrester, which now produces a smaller discharge voltage. It is this voltage that is impressed upon the cable.

To further enhance the protection, the first span on either side of the riser pole can be shielded to prevent direct strokes to the line.

Recently there has been renewed interest in the scheme.[12] There is empirical evidence that the scout scheme helps prevent open-point failures of both cable and transformers. The expense of changing out a transformer far exceeds the additional cost of the scout arresters. Simulations suggest that while the nominal arrester discharge voltage may be reduced only a few percent, the greatest benefit of the scout scheme may be that it greatly reduces the rate-of-rise of surge voltages entering the cable. These steep-fronted surges reflect off the open point and frequently cause failures at the first or second pad-mounted transformer from the end. Due to lead lengths, arresters are not always effective against such impulses. The scout scheme practically eliminates these from the cable.

Many distribution feeders in densely populated areas will have scout schemes by default. There are sufficient numbers of transformers that there are already arresters on either side of the riser pole.

4.6 Load-Switching Transient Problems

This section describes some transients problems related to load switching.

4.6.1 Nuisance tripping of ASDs

Most adjustable-speed drives typically use a voltage source inverter (VSI) design with a capacitor in the dc link. The controls are sensitive to dc overvoltages and may trip the drive at a level as low as 117 percent. Since transient voltages due to utility capacitor switching typically exceed 130 percent, the probability of nuisance tripping of the drive is high. One set of typical waveforms for this phenomenon is shown in Fig. 4.28.

The most effective way to eliminate nuisance tripping of small drives is to isolate them from the power system with ac line chokes. The additional series inductance of the choke reduces the transient voltage magnitude that appears at the input to the ASD. Determining the precise inductor size required for a particular application (based on utility capacitor size, transformer size, etc.) requires a fairly detailed transient simulation. A series choke size of 3 percent based on the drive kVA rating is usually sufficient.

4.6.2 Transients from load switching

Deenergizing inductive circuits with air-gap switches, such as relays and contactors, can generate bursts of high-frequency impulses. Figure 4.29 shows an example. ANSI/IEEE C62.41 cites a representative 15-ms burst composed of impulses having 5-ns rise times and 50-ns durations. There is very little energy in these types of transient due to their short duration, but they can interfere with the operation of electronic loads.

Such electrical fast transient (EFT) activity, producing spikes up to 1 kV, is frequently due to cycling motors, such as air conditioners and elevators. Transients as high as 3 kV can be caused by operation of arc welders and motor starters.

The duration of each impulse is short compared to the travel time of building wiring, thus the propagation of these impulses through the wiring can be analyzed with traveling-wave theory. The impulses attenuate very quickly as they propagate through a building. Therefore, in most cases, the only protection needed is electrical separation. Physical separation is also required be-

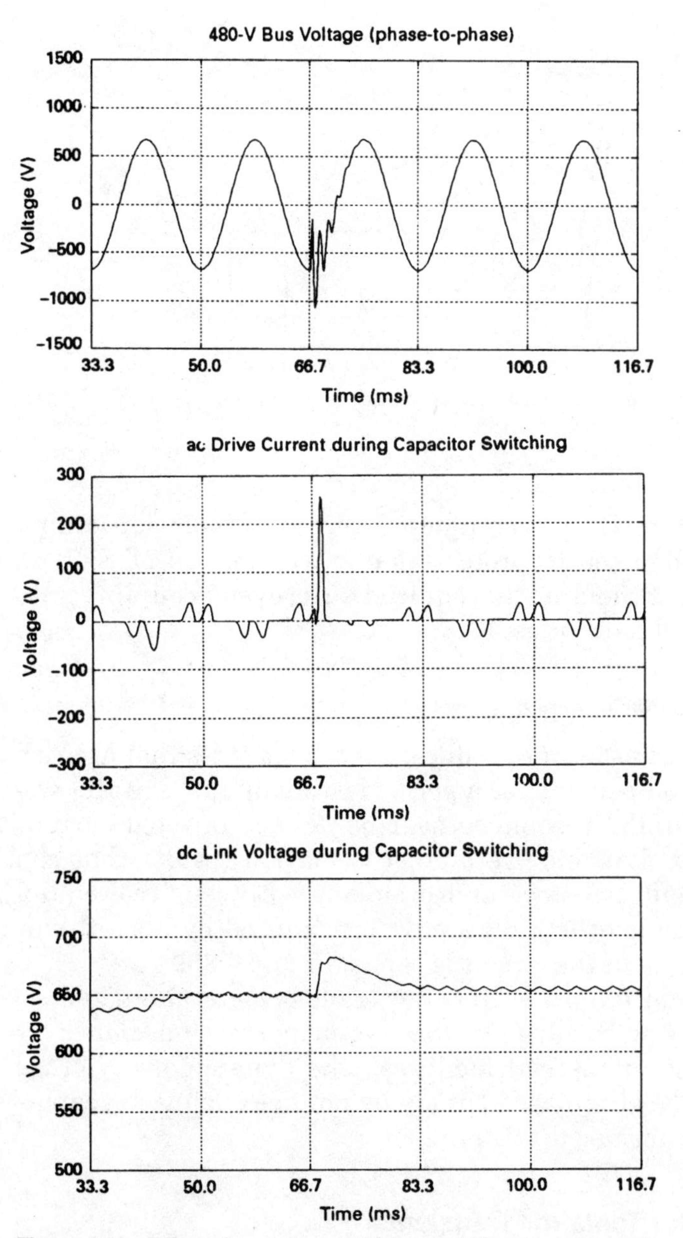

Figure 4.28 Effect of capacitor switching on ASD ac current and dc voltage.

cause the high rate of rise allows these transients to couple into nearby sensitive equipment.

EFT suppression may be required with extremely sensitive equipment in close proximity to a disturbing load, such as a

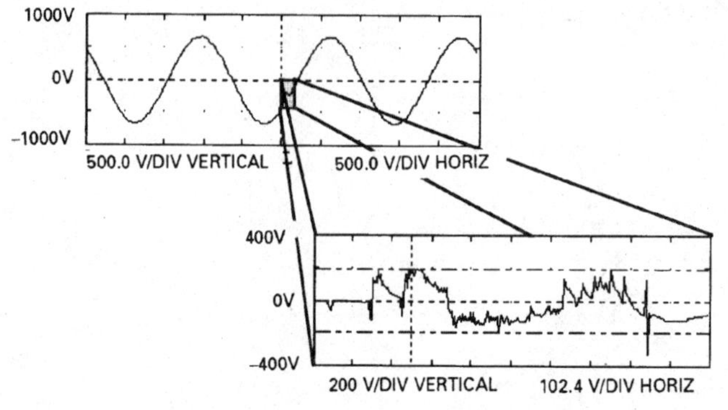

Figure 4.29 Fast transients caused by deenergizing an inductive load.

computer room. High-frequency filters and isolation transformers can be used to protect against conduction of EFT on power cables. Shielding is required to prevent coupling into equipment and data lines.

4.6.3 Transformer energizing

Energizing a transformer produces inrush currents that are rich in harmonic components for a period lasting up to 1 s. If the system has a parallel resonance near one of the injected current frequencies, a dynamic overvoltage condition results that can cause failure of arresters and problems with sensitive equipment. A dynamic overvoltage waveform caused by a third-harmonic resonance in the circuit is shown in Fig. 4.30.

This problem occurs with large transformers energized simultaneously with large power factor correction capacitor banks in large industrial facilities. The dynamic overvoltage problem can be eliminated simply by not energizing the capacitor and transformer together.

4.7 Computer Tools for Transients Analysis

The most widely used computer program for transients analysis of power systems is the Electromagnetic Transients Program, commonly known as EMTP. It was originally developed by Hermann W. Dommel at the Bonneville Power Administration (BPA) in the late 1960s[13] and has been continuously upgraded

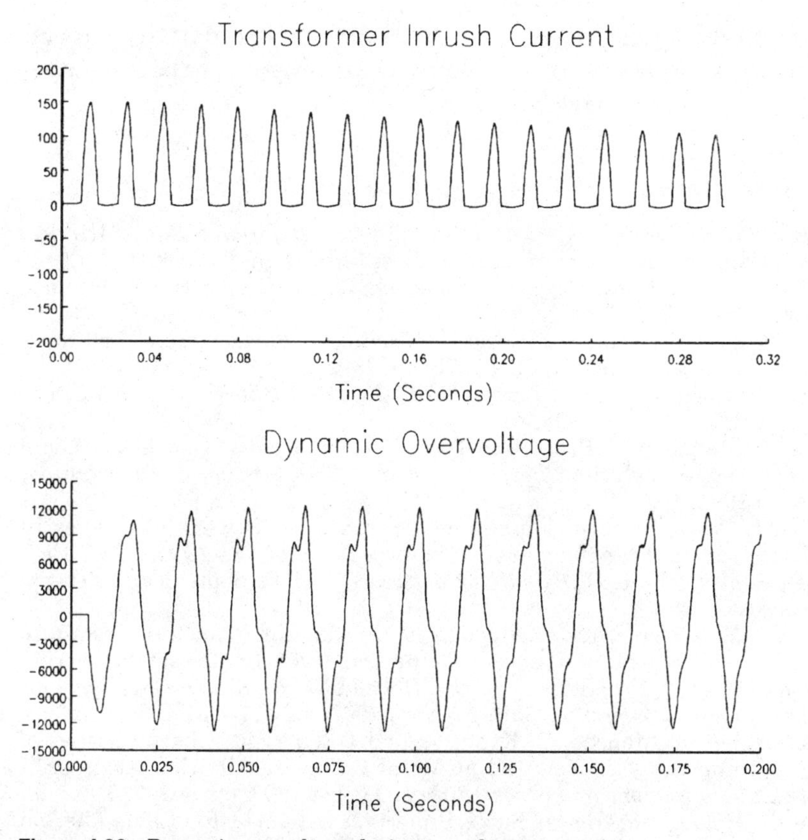

Figure 4.30 Dynamic overvoltage during transformer energizing.

since. One group currently performing development and support is coordinated by EPRI and the Development Coordination Group (DCG). Another version of the program is the Alternate Transients Program (ATP), applicable to personal computers (PCs). It is available through the Can/Am EMTP User's Group, organized by W. Scott Meyer, who has had a leading role in the development and maintenance of the program for many years.

Another commercial analysis tool is the PSCAD™/EMTDC™ program developed by the Manitoba HVDC Research Center. This program features a very sophisticated graphical user interface that allows the user to assemble the circuit and observe its behavior while the solution is proceeding.

There are numerous computer programs developed for analysis of electronic circuits that can also be adapted to analyze power systems. One example is the well-known SPICE program[14] and its derivatives. These programs are usually less efficient in power systems problems than programs that are specially designed for power systems.

4.8 References

1. *Electrical Transmission and Distribution Reference Book,* 4th ed., Westinghouse Electric Corporation, East Pittsburgh, Pa., 1964.
2. *Electrical Distribution-System Protection,* 3d ed., Cooper Power Systems, Franksville, Wis., 1990.
3. K. Berger, R. B. Anderson, and H. Kroninger, "Parameters of Lightning Flashes," *Electra,* no. 41, July 1975, pp. 23–27.
4. R. Morrison and W. H. Lewis, *Grounding and Shielding in Facilities,* John Wiley & Sons, 1990.
5. R. A. Stansberry, "Protecting Distribution Circuits: Overhead Shield Wire Versus Lightning Surge Arresters," *Transmission & Distribution,* April 1991, pp. 56*ff.*
6. IEEE Transformers Committee, "Secondary (Low-Side) Surges in Distribution Transformers," in *Proceedings of the 1991 IEEE PES Transmission and Distribution Conference,* Dallas, September 1991, pp. 998–1008.
7. C. W. Plummer et al., "Reduction in Distribution Transformer Failure Rates and Nuisance Outages Using Improved Lightning Protection Concepts," in *Proceedings of the 1994 IEEE PES Transmission and Distribution Conference,* Chicago, April 1994, pp. 411–416.
8. P. Barker, R. Mancao, D. Kvaltine, and D. Parrish, "Characteristics of Lightning Surges Measured at Metal Oxide Distribution Arresters," *IEEE Transactions on Power Delivery,* October 1993, pp. 301–310.
9. R. H. Hopkinson, "Better Surge Protection Extends URD Cable Life," in *Proceedings of the 1984 IEEE/PES T&D Conference and Exposition,* Kansas City, Mo.
10. G. L. Goedde, R. C. Dugan, and L. D. Rowe, "Full Scale Lightning Surge Tests of Distribution Transformers and Secondary Systems," in *Proceedings of the 1991 IEEE PES Transmission and Distribution Conference,* Dallas, September 1991, pp. 691–697.
11. S. S. Kershaw, Jr., "Surge Protection for High Voltage Underground Distribution Circuits," in *Conference Record of the IEEE Conference on Underground Distribution,* Detroit, September 1971, pp. 370–384.
12. M. B. Marz, T. E. Royster, and C. M. Wahlgren, "A Utility's Approach to the Application of Scout Arresters for Overvoltage Protection of Underground Distribution Circuits," in *1994 IEEE Transmission and Distribution Conference Record,* Chicago, April 1994, pp. 417–425.
13. H. W. Dommel, "Digital Computer Solution of Electromagnetic Transients in Single and Multiphase Networks," *IEEE Transactions on Power Apparatus and Systems,* vol. PAS-88, April 1969, pp. 388–399.
14. L. W. Nagel, "SPICE2: A Computer Program to Simulate Semiconductor Circuits," Ph.D. thesis, University of California, Berkeley, Electronics Research Laboratory, No. ERL-M520, May 1975.
15. IEEE Standard C62.41-1991, *IEEE Recommended Practice on Surge Voltages in Low-Voltage AC Power Circuits*, Piscataway, N.J., 1991.

Chapter

5

Harmonics

A good assumption for most utilities in the United States is that the sine wave voltage generated in bulk power stations is very good. In most areas, the voltage found on transmission systems typically has much less than 1.0 percent distortion. However, as we move closer to the load, the distortion increases. At some loads, the current waveforms will barely resemble a sine wave. Electronic power converters can chop the current into seemingly arbitrary waveforms. While there are a few cases where the distortion is random, most distortion is periodic, or harmonic. That is, it is nearly the same cycle after cycle, changing very slowly, if at all. This has given rise to the widespread use of the term *harmonics* to describe perturbations in the waveform. As we shall see, this term must be carefully qualified to make sense and we will endeavor in this chapter to remove some of the mystery of harmonics in power systems.

When electronic power converters first became commonplace in the late 1970s, many utility engineers became quite concerned about the ability of the power system to accommodate the harmonic distortion. Many dire predictions were made about the fate of power systems if these devices were permitted to exist. While some of these concerns were probably overstated, the field of power quality analysis owes a great debt of gratitude to these people because their concern over this "new" problem of harmonics sparked the research that has eventually led to much of the knowledge about all aspects of power quality.

To some, harmonic distortion is still the most significant power quality problem. It is not hard to understand how an engineer faced with a difficult harmonics problem can come to hold that opinion. Harmonics problems counter many of the conventional rules of power system design and operation that consider only the fundamental frequency. Therefore, the engineer is faced with unfamiliar phenomena that require unfamiliar tools to analyze and unfamiliar equipment to solve. Although harmonics problems can be difficult, they are not actually very numerous on utility systems. Only a few percent of utility distribution feeders in the United States have a sufficiently severe harmonics problem to require attention. In contrast, voltage sags and interruptions are nearly universal to every feeder and represent the most numerous and significant power quality deviations. The end-user sector suffers more from harmonic problems than the utility sector. Industrial users with adjustable-speed drives, arc furnaces, induction furnaces, and the like, are much more susceptible to problems stemming from harmonic distortion.

Harmonic distortion is not a new phenomenon on power systems. Concern over distortion has ebbed and flowed a number of times during the history of alternating current electric power systems. Scanning the technical literature of the 1930s and 1940s, one notices many articles on the subject. Then the primary sources were the transformers and the primary problem was inductive interference with open-wire telephone systems. The forerunners of modern arc lighting were being introduced and were causing quite a stir because of their harmonic content—not unlike the stir caused by electronic power converters in more recent times.

Fortunately, we have found over the years that if the system is properly sized to handle the power demands of the load, there is a low probability that harmonics will cause a problem with the power system, although they may cause problems with telecommunications. The power system problems arise most frequently when the capacitance in the system results in resonance at a critical harmonic frequency that dramatically increases the distortion above normal amounts. While these problems occur on utility systems, the most severe cases are usually found in industrial power systems because of the higher degree of resonance achieved.

5.1 Harmonic Distortion

Harmonic distortion is caused by *nonlinear* devices in the power system. A nonlinear device is one in which the current is

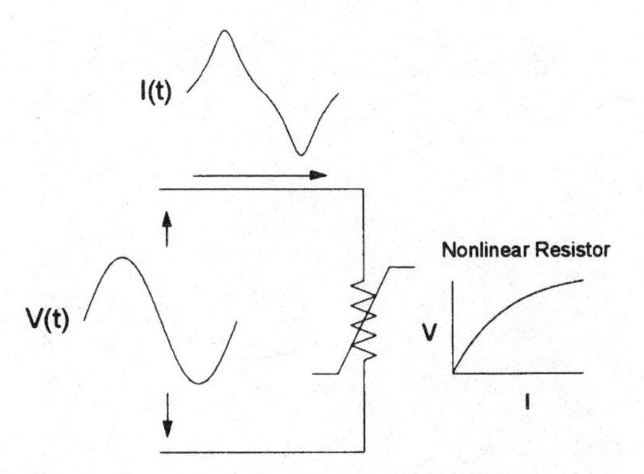

Figure 5.1 Current distortion caused by nonlinear resistance.

not proportional to the applied voltage. Figure 5.1 illustrates this concept by the case of a sinusoidal voltage applied to a simple nonlinear resistor in which the voltage and current vary according to the curve shown. While the applied voltage is perfectly sinusoidal, the resulting current is distorted. Increasing the voltage by a few percent may cause the current to double and take on a different waveshape. This, in essence, is the source of harmonic distortion in the power system.

Figure 5.2 illustrates that any periodic, distorted waveform can be expressed as a sum of sinusoids. That is, when the waveform is identical from one cycle to the next, it can be represented as a sum of pure sine waves in which the frequency of each sinusoid is an integer multiple of the fundamental frequency of the distorted wave. This multiple is called a *harmonic* of the fundamental, hence the name of this subject matter. The sum of sinusoids is referred to as a *Fourier series* after the great mathematician who discovered the concept.

The advantage of using a Fourier series to represent distorted waveforms is that it is much easier to find the system response to an input that is sinusoidal. Conventional steady-state analysis techniques can be used. The system is analyzed separately at each harmonic. Then the outputs at each frequency are combined to form a new Fourier series, from which the output waveform may be computed, if desired. Often, only the magnitudes of the harmonics are of interest.

When both the positive and negative half-cycles of a waveform have identical shapes, the Fourier series contains only *odd* harmonics. This offers a further simplification for most power system studies because most common harmonic-producing devices

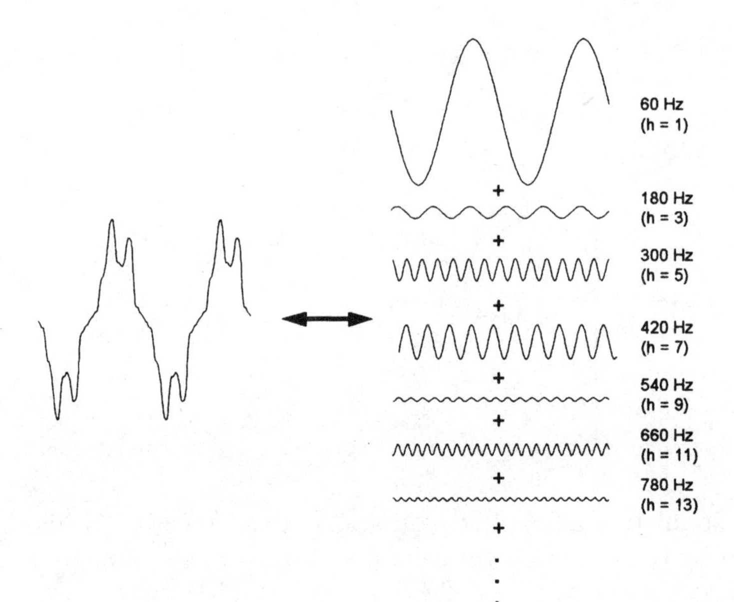

Figure 5.2 Fourier series representation of a distorted waveform.

look the same to both polarities. In fact, the presence of even harmonics is often a clue that there is something wrong—either with the load equipment or with the transducer used to make the measurement. There are notable exceptions to this such as half-wave rectifiers and arc furnaces when the arc is random.

Usually, the higher-order harmonics (above the range of the 25th to 50th, depending on the system) are negligible for power system analysis. While they may cause interference with low-power electronic devices, they are usually not damaging to the power system. It is also difficult to collect sufficiently accurate data to model power systems at these frequencies.

If we break the power system into series and shunt elements, as is conventional practice, the vast majority of the nonlinearities in the system are found in *shunt* elements (i.e., loads). The series impedance of the power delivery system (i.e., the short circuit impedance between the source and the load) is remarkably linear. In transformers, also, the source of harmonics is the shunt branch (magnetizing impedance) of the common "T" model; the leakage impedance is linear. Thus, the main sources of harmonic distortion will ultimately be end-user loads. This is not to say that all end users who experience harmonic distortion will themselves have significant sources of harmonics, but that the harmonic distortion generally originates with some end-user's load or combination of loads.

5.2 Voltage vs. Current Distortion

The word *harmonics* is often used by itself without further qualification. For example, it is common to hear that an adjustable-speed drive or an induction furnace can't operate properly because of harmonics. What does that mean? It could mean one of three things:

1. The harmonic voltages are too great (the voltage too distorted) for the the control to properly determine firing angles.
2. The harmonic currents are too great for the capacity of some device in the power supply system such as a transformer and the machine must be operated at a lower than rated power.
3. The harmonic voltages are too great because the harmonic currents produced by the device are too great for the given system condition.

As suggested by this list, there are separate causes and effects for voltages and currents as well as some relationship between them. Thus, the term *harmonics* by itself is too ambiguous to definitively describe a problem.

Nonlinear loads appear to be sources of harmonic current in shunt with and injecting harmonic currents *into* the power system. For nearly all analyses, it is sufficient to treat these harmonic-producing loads simply as current sources. There are exceptions to this that we will describe later.

As Fig. 5.3 shows, voltage distortion is the result of distorted currents passing through the linear, series impedance of the power delivery system. Although we have assumed here that the source bus contains only fundamental frequency voltage, the harmonic currents passing through the impedance of the system cause a voltage drop for each harmonic. This results in voltage harmonics appearing at the load bus. The amount of voltage distortion depends on the impedance and the current. Assuming the load bus distortion stays within reasonable limits (e.g., less than 5 percent), the amount of harmonic current produced by the load is nearly constant for each load level.

While the load current harmonics ultimately cause the voltage distortion, it should be noted that load has no control over

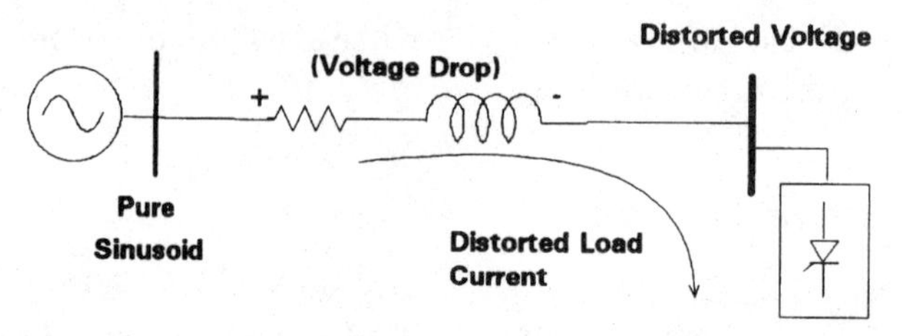

Figure 5.3 Harmonic currents flowing through the system impedance results in harmonic voltages at the load.

the voltage distortion. The same load in two different locations on the power system will result in two different voltage distortion values. Recognition of this fact is the basis for the division of responsibilities for harmonic control that is found in standards such as IEEE Standard 519-1992:

1. The control over the amount of harmonic current injected into the system takes place at the end-use application.
2. Assuming the harmonic current injection is within reasonable limits, the control over the voltage distortion is exercised by the entity having control over the system impedance, which is often the utility.

One must be careful when describing harmonic phenomena to understand that there are distinct differences between the causes and effects of harmonic voltages and currents. The use of the term *harmonics* should be qualified accordingly. By popular convention in the power industry, the majority of times the term is used by itself when referring to load apparatus, the speaker is referring to the harmonic currents. When referring to the utility system, the voltages are generally the subject. To be safe, make a habit of asking for clarification.

5.3 Harmonics vs. Transients

Harmonic distortion is blamed for many power quality disturbances that are actually transients. A measurement of the event may show a distorted waveform with obvious high-frequency components. Although transient disturbances contain high-frequency components, transients and harmonics are distinctly different phenomena and are analyzed differently. Transient waveforms exhibit the high frequencies only briefly after there has been an abrupt change in the power system. The frequencies are not necessarily harmonics; they are whatever the natural frequencies of the system are at the time of the switching operation. These frequencies have no relation to the system fundamental frequency.

Harmonics, by definition, occur in the steady state, and are integer multiples of the fundamental frequency. The waveform distortion that produces the harmonics is present continually, or at least for several seconds. Transients are usually dissipated within a few cycles. Transients are associated with changes in the system such as switching a capacitor bank. Harmonics are associated with the continuing operation of a load.

One case in which the distinction is blurred is transformer energization. This is a transient event, but can produce considerable waveform distortion for many seconds and has been known to excite system resonances.

5.4 Total Harmonic Distortion and rms Value

There are several measures commonly used for indicating the harmonic content of a waveform with a single number. One of the most common is total harmonic distortion (THD), which can be calculated for either voltage or current:

$$\mathrm{THD} = \frac{\sqrt{\sum_{h=2}^{h_{\max}} M_h^2}}{M_1} \tag{5.1}$$

where M_h is the rms value of harmonic component h of the quantity M. THD is a measure of the *effective value* of the harmonic components of a distorted waveform, that is, the potential heating value of the harmonics relative to the fundamental.

The rms value of the total waveform is not the sum of the individual components, but is the square root of the sum of the squares. THD is related to the rms value of the waveform as follows:

$$\mathrm{rms} = \sqrt{\sum_{h=1}^{h_{\max}} M_h^2} = M_1 \cdot \sqrt{1 + \mathrm{THD}^2} \tag{5.2}$$

THD is a very useful quantity for many applications, but its limitations must be realized. It can provide a good idea of how much extra heat will be realized when a distorted voltage is applied across a resistive load. Likewise, it can give an indication of the addition losses caused by the current flowing through a conductor. However, it is not a good indicator of the voltage stress within a capacitor because that is related to the peak value of the voltage waveform, not its heating value.

Harmonic voltages are almost always referenced to the fundamental value of the waveform at the time of the sample. Because voltage varies only a few percent, the voltage THD is nearly always a meaningful number. This is not the case for current. A small current may have a high THD but not be a significant threat to the system. Since most monitoring devices report THD based on the present sample, the user may be misled into thinking the current is dangerous. Some analysts have attempted to avoid this difficulty by referring THD to the fundamental of the peak demand current rather than the fundamental of the present sample. This is called total demand distortion or, simply, TDD, and serves as the basis for the guidelines in IEEE Standard 519-1992.

5.5 Power and Power Factor

Harmonic distortion complicates the computation of power and power factor because many of the simplifications power engineers use for power frequency analysis do not apply.

There are three standard quantities associated with power:

Apparent power, S. The product of the rms voltage and current.

Active power, P. The average rate of delivery of energy.

Reactive power, Q. The portion of the apparent power that is out of phase, or in quadrature, with the active power.

At fundamental frequency, it is common to relate these quantities as follows:

$$P = S \cos \theta \tag{5.3}$$

$$Q = S \sin \theta \tag{5.4}$$

where θ = phase angle between voltage and current

The factor cos θ is commonly called the *power factor*. However, a more correct definition is to simply define the power factor (PF) as

$$\text{PF} = \frac{P}{S} \tag{5.5}$$

S and P are unambiguously defined even with distorted voltage and current, while there is no clear concept of phase angle that applies to the multiple-frequency situation:

$$S = V_{\text{rms}\,131} I_{\text{rms}} \tag{5.6}$$

$$P = \frac{1}{T}\int_0^T v(t)i(t)dt \tag{5.7}$$

When the voltage, V, is entirely fundamental frequency, P resolves to the familiar form

$$P = \frac{V_1 I_1}{2}\cos\theta_1 = V_{1\text{rms}} I_{1\text{rms}} \cos\theta_1 \tag{5.8}$$

which indicates that the average active power is a function only of the fundamental frequency quantities. Because the voltage distortion is generally very low on power systems (less than 5 percent), this is a good approximation regardless of how distorted the current is.

On the other hand, the apparent power and reactive power terms are greatly influenced by the distortion. The apparent power, S, is a measure of the potential impact of the load on the thermal capability of the system. It is proportional to the rms of the distorted current and it computation is straightforward, although slightly more complicated than the sinusoidal case. Also, many current probes can now directly report the true rms value of a distorted waveform.

There is some disagreement among harmonics analysts on how to define Q in the presence of harmonic distortion. If it were not for the fact that many utilities meter Q and compute demand billing from the power factor computed by Q, it might be a moot point. It is more important to determine P and S; P defines how much energy is being consumed while S defines the capacity of the power system required to deliver P. Q is not actually very useful by itself.

The reactive power when distortion is present has another interesting peculiarity. In fact, it may not be appropriate to call it reactive *power.* The concept of var flow in the power system is deeply ingrained in the minds of most power engineers. What many do not realize is that this concept is valid only in the sinusoidal steady state. When distortion is present, the component of S that remains after P is taken out is not conserved—i.e., it does not sum to zero at a node. Power quantities are presumed to flow around the system in a conservative manner.

This does not imply that P is not conserved or that current is not conserved because the conservation of energy and Kirchoff's current laws are still applicable for any waveform. The reactive components actually sum in quadrature (square root of the sum of the squares). This has prompted some analysts to propose that Q be used to denote the reactive components that are conserved and introduce a new quantity for the components that are not. Many call this quantity D, for *distortion power,* or, simply, *distortion voltamperes.*[1] It has units of voltamperes, but it may not be strictly appropriate to refer to this quantity as *power,* because it does not flow through the system as power is assumed to do. In this concept, Q consists of the sum of the traditional reactive power values at each frequency. D represents all cross-products of voltage and current at different frequencies, which yield no average power. P, Q, D, and S are related as follows, using the definitions for S and P above as a starting point:

$$S = \sqrt{P^2 + Q^2 + D^2} \tag{5.9}$$

$$Q = \sum_{k} V_k I_k \sin \theta_k \tag{5.10}$$

Therefore, D can be determined after S, P, and Q by:

$$D = \sqrt{S^2 - P^2 - Q^2} \tag{5.11}$$

Some prefer to use a three-dimensional vector chart to demonstrate the relationships of the components as shown in Fig. 5.4. P and Q contribute the traditional sinusoidal components to S, while D represents the additional contribution to the apparent power by the harmonics.

The fundamental frequency component of the reactive power, Q_1, is useful for helping engineers size capacitors for power factor correction. Capacitors can only correct for Q_1. The term *dis-*

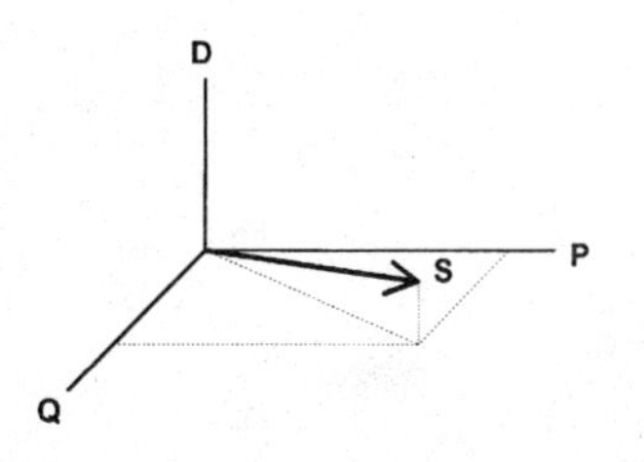

Figure 5.4 Relationship of components of the apparent power.

placement power factor is used to describe the power factor using the fundamental frequency components only. Power quality monitoring instruments now commonly report this quantity as well as the *true power factor,* which is the same quantity defined as PF previously [Eq. (5.5)]. Many devices such as switch-mode power supplies and pulse-width modulated (PWM) adjustable-speed drives have a near-unity displacement power factor, but the true power factor may be 0.5 to 0.6. An ac-side capacitor will do little to improve the true power factor in this case. In fact, if it results in resonance, the distortion may increase, causing the power factor to degrade. The true power factor indicates how large the power delivery system must be built to supply a given load. In this example, using only the displacement power factor would give a false sense of security that all is well.

Many demand metering devices will record only Q_1. Fortunately, in most cases, the current at the metering point is not as greatly distorted as individual load currents and the error is small (and in the customer's favor). There are some exceptions to this such as pumping stations where a PWM drive is the only load on the meter. While the energy meter should be sufficiently accurate given that the voltage has low distortion, the demand metering could have substantial error.

The bottom line is that distortion results in additional current components flowing in the system that do not yield any net energy except that they cause losses in the power system elements they pass through. This requires the system to be built to a slightly larger capacity to deliver the power to the load.

5.6 Triplen Harmonics

Triplen harmonics are the odd multiples of the third harmonic (h = 3, 9, 15, 21, ...). They deserve special consideration because the system response is often considerably different for

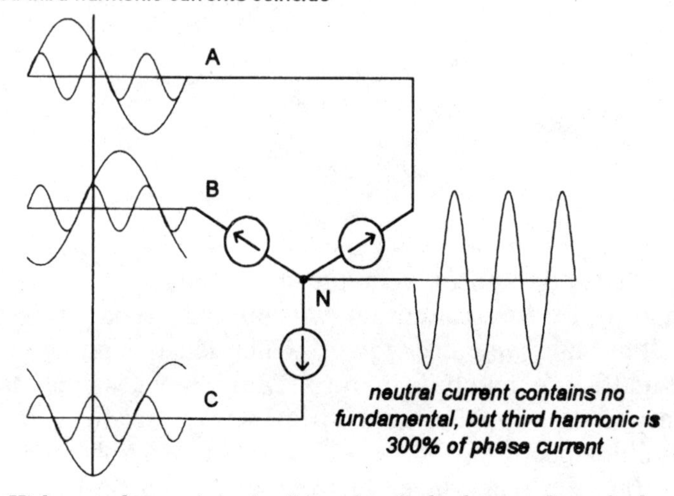

Figure 5.5 High neutral currents in circuits serving single-phase nonlinear loads.

triplens than for the rest of the harmonics. Triplens become an important issue for grounded-wye systems with current flowing on the neutral. Two typical problems are overloading the neutral and telephone interference. One also hears occasionally of devices that misoperate because the line-to-neutral voltage is badly distorted by the triplen harmonic voltage drop in the neutral conductor.

For the system of perfectly balanced single-phase loads illustrated in Fig. 5.5, we assume that fundamental and third harmonic components are present. Summing the currents at node N, the fundamental current components in the neutral are found to be zero, but the third-harmonic components are three times the third-harmonic-phase currents because they naturally coincide in phase and time.

Transformer winding connections have a significant impact on the flow of triplen harmonic currents from single-phase nonlinear loads. Two cases are shown in Fig. 5.6. In the wye-delta transformer (top), the triplen harmonic currents are shown entering the wye side. Since they are in phase, they add in the neutral. The delta winding provides ampere-turn balance so that they can flow, but they remain trapped in the delta and do not show up in the line currents on the delta side. When the currents are balanced, the triplen harmonic currents behave exactly as zero-sequence currents, which is precisely what they

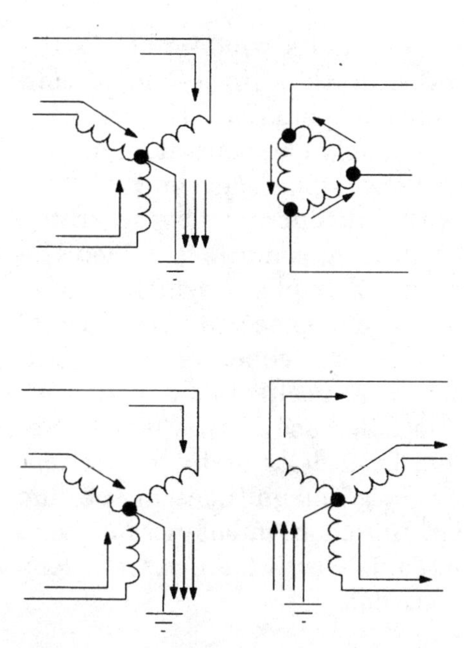

Figure 5.6 Flow of third-harmonic current in three-phase transformers.

are. This type of transformer connection is the most common employed in utility distribution substations with the delta winding connected to the transmission feed.

Using grounded-wye windings on both sides of the transformer (bottom) allows balanced triplens to flow from the low-voltage system to the high-voltage system unimpeded. They will be present in equal proportion on both sides. Many loads in the United States are served in this fashion.

Some important implications of this related to power quality analysis are:

1. Transformers, particularly the neutral connections, are susceptible to overheating when serving single-phase loads on the wye side that have high third-harmonic content.
2. Measuring the current on the delta side of a transformer will not show the triplens and, therefore, not give a true idea of the heating the transformer is being subjected to.
3. The flow of triplen harmonic currents can be interrupted by the appropriate isolation transformer connection.

Removing the neutral connection in one or both wye windings blocks the flow of triplen harmonic current. There is no place for ampere-turn balance. Likewise, a delta winding blocks the flow from the line. One should note that three-legged core transformers behave as if they have a "phantom" delta tertiary winding. Therefore, a wye-wye with only one neutral point connected will still be able to conduct the triplen harmonics.

These rules about triplen harmonic current flow in transformers apply only to *balanced* loading conditions. When the phases are not balanced, currents of triplen harmonics may very well show up where they are not expected. The normal mode for triplen harmonics is to be zero sequence. During imbalances, triplen harmonics may have positive or negative sequence components, too. One notable case of this is a three-phase arc furnace. Although fed by a delta-delta connection, the third harmonics show up in large magnitudes in the line current when the furnace is operating in an imbalanced state.

But to the extent that the system is *mostly* balanced, triplens *mostly* behave in the manner described.

5.7 Single-Phase Power Supplies

Electronic power converter loads with their capacity for producing harmonic currents now constitute the most important class of nonlinear loads in the power system. Advances in semiconductor device technology have fueled a revolution in power electronics over the past decade, and there is every indication that this trend will continue. Equipment includes adjustable-speed motor drives, electronic power supplies, dc motor drives, battery chargers, electronic ballasts, and many other rectifier/inverter applications.

A major harmonics concern in commercial buildings is that power supplies for single-phase electronic equipment will produce too much distortion for the wiring. Direct current power for modern electronic and microprocessor-based office equipment is commonly derived from single-phase full-wave diode bridge rectifiers. The percentage of load which contains electronic power supplies is increasing at a dramatic pace, with the increased utilization of personal computers in every commercial sector.

Two major types of single-phase power supplies are common. Older technologies use ac-side voltage control methods, such as transformers, to reduce voltages to the level required for the dc bus. The inductance of the transformer provides a beneficial

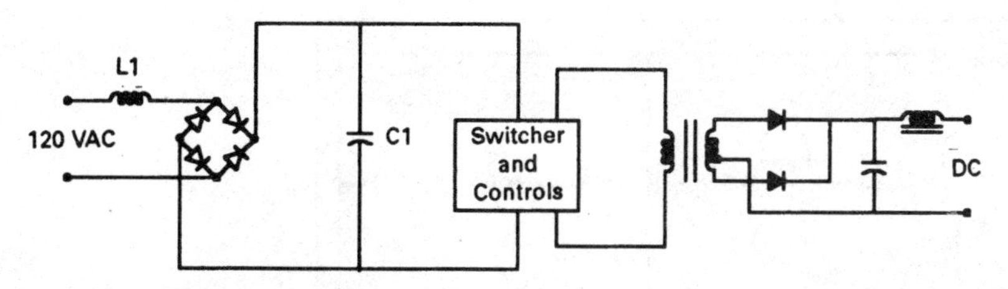

Figure 5.7 Switch-mode power supply.

side effect by smoothing the input current waveform, reducing harmonic content. Newer technology, switch-mode power supplies (Fig. 5.7), use dc/dc conversion techniques to achieve a smooth dc output with small, lightweight components. The input diode bridge is directly connected to the ac line, eliminating the transformer. This results in a coarsely regulated dc voltage on the capacitor. This dc is then converted back to ac at a very high frequency by the switcher and subsequently rectified again. Personal computers, printers, copiers, and most other single-phase electronic equipment now almost universally employ switch-mode power supplies. The key advantages are the light weight, compact size, efficient operation, and lack of need for a transformer. They can usually tolerate large variations in input voltage.

Because there is no large ac-side inductance, input current to the power supply comes in very short pulses as the capacitor, C_1, regains its charge on each half cycle. Figure 5.8 illustrates the current waveform and spectrum for an entire circuit supplying a variety of electronic equipment with switch-mode power supplies (SMPS).

A distinctive characteristic of switch-mode power supplies is a very high third-harmonic content in the current. Since third-harmonic current components are additive in the neutral of a three-phase system, the increasing application of switch-mode power supplies causes concern for overloading of neutral conductors, especially in older buildings where an undersized neutral may have been installed. Concern for transformer heating is also important when the load includes a significant amount of switch-mode power supplies.

Switched-mode power supplies are also beginning to find application in electronic ballasts for fluorescent lighting systems. The high-frequency, controlled-output voltage that is possible with transistorized inverters increases fluorescent tube effi-

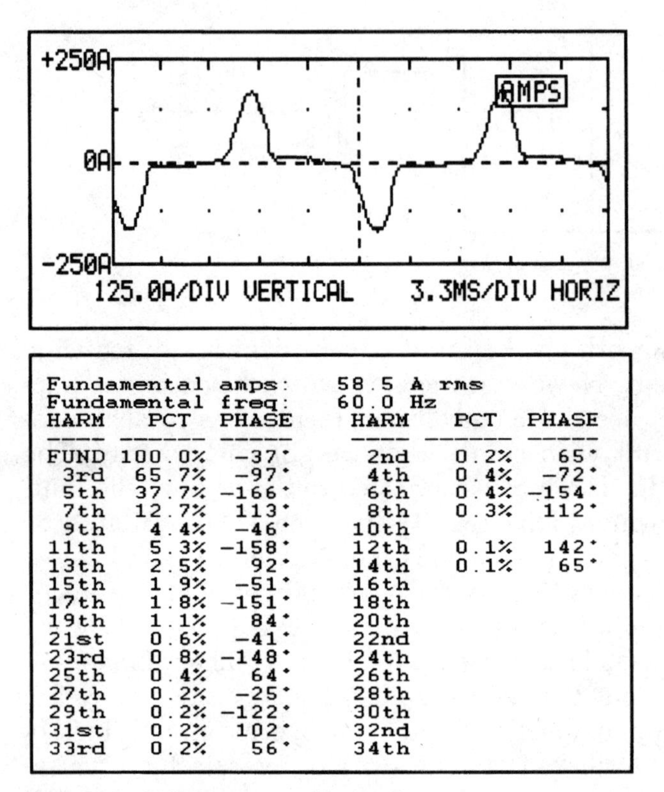

Fundamental amps: 58.5 A rms
Fundamental freq: 60.0 Hz

HARM	PCT	PHASE	HARM	PCT	PHASE
FUND	100.0%	−37°	2nd	0.2%	65°
3rd	65.7%	−97°	4th	0.4%	−72°
5th	37.7%	−166°	6th	0.4%	−154°
7th	12.7%	113°	8th	0.3%	112°
9th	4.4%	−46°	10th		
11th	5.3%	−158°	12th	0.1%	142°
13th	2.5%	92°	14th	0.1%	65°
15th	1.9%	−51°	16th		
17th	1.8%	−151°	18th		
19th	1.1%	84°	20th		
21st	0.6%	−41°	22nd		
23rd	0.8%	−148°	24th		
25th	0.4%	64°	26th		
27th	0.2%	−25°	28th		
29th	0.2%	−122°	30th		
31st	0.2%	102°	32nd		
33rd	0.2%	56°	34th		

Figure 5.8 SMPS current and harmonic spectrum.

ciency, and permits more sophisticated control, such as dimming. The harmonic current injected by many electronic ballasts looks very similar to power supplies used in computers and other electronic equipment. Increased harmonic generation from fluorescent lighting can be very important because lighting typically accounts for 40 to 60 percent of a commercial building load. Some vendors have responded with designs that produce a much cleaner waveform.

5.8 Three-Phase Power Converters

Three-phase electronic power converters differ from single-phase converters mainly because they do not generate third-harmonic currents. This is a great advantage because that is the largest component. However, they can still be significant sources of harmonics at their characteristic frequencies, as shown in Fig. 5.9. This is a typical current source type of ASD.

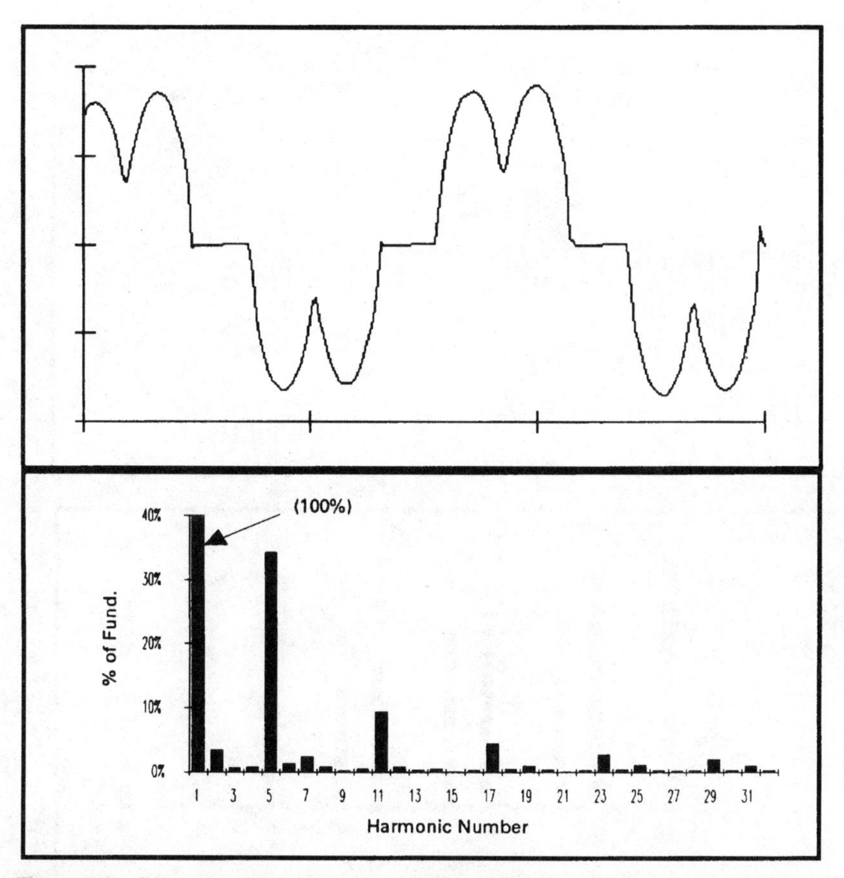

Figure 5.9 Current and harmonic spectrum for CSI-type ASD.

The harmonic spectrum given in Fig. 5.9 would also be typical of a dc motor drive input current. Voltage source inverter drives [such as pulse-width modulated (PWM) type drives] can have much higher distortion levels as shown in Fig. 5.10.

The input to the PWM drive is generally designed like a three-phase version of the switch-mode power supply in computers. The rectifier feeds directly from the ac bus to a large capacitor on the dc bus. With little intentional inductance, the capacitor is charged in very short pulses, creating the distinctive "rabbit ear" ac-side current waveform with very high distortion. Whereas the switch-mode power supplies are generally for very small loads, PWM drives are now being applied for loads up to 500 hp. This is a justifiable cause for concern from power engineers.

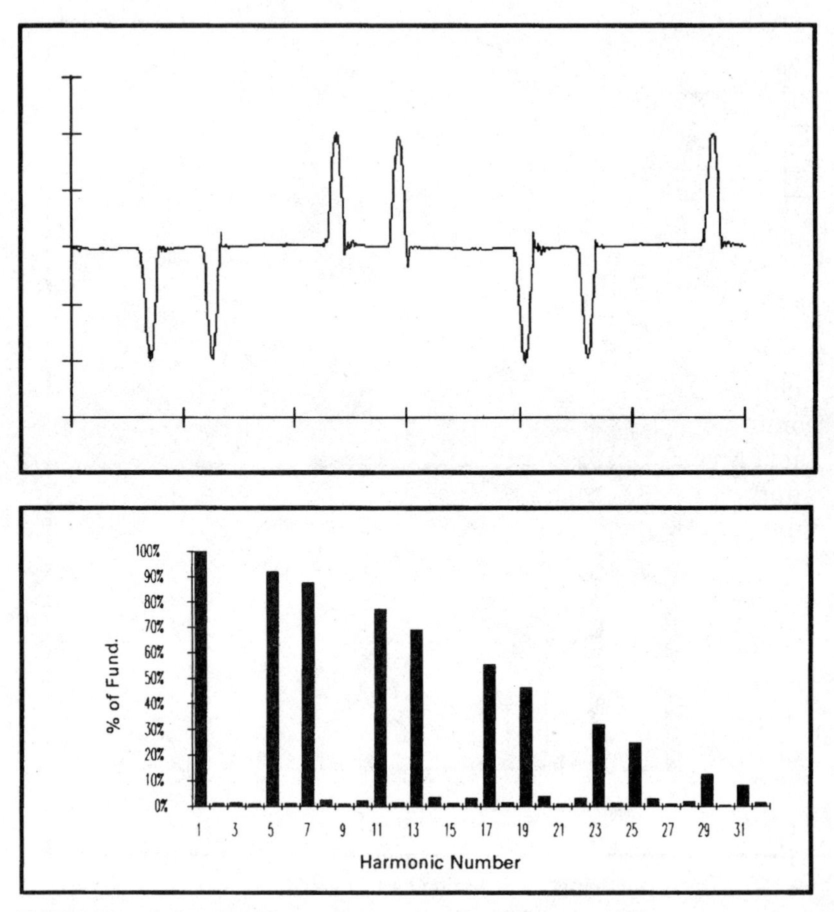

Figure 5.10 Current and harmonic spectrum for PWM-type ASD.

5.8.1 dc drives

Rectification is the only step required for dc drives. Therefore, they have the advantage of relatively simple control systems. Compared with ac drive systems, the dc drive offers a wider speed range and higher starting torque. However, purchase and maintenance costs for dc motors are high, while the cost of power electronic devices has been dropping year after year. Thus, economic considerations limit the dc drive to applications that require the speed and torque characteristics of the dc motor.

Most dc drives use the six-pulse rectifier shown in Fig. 5.11. Large drives may employ a 12-pulse rectifier. This reduces thyristor current duties, and reduces some of the larger ac current harmonics. The two largest harmonic currents for the six-

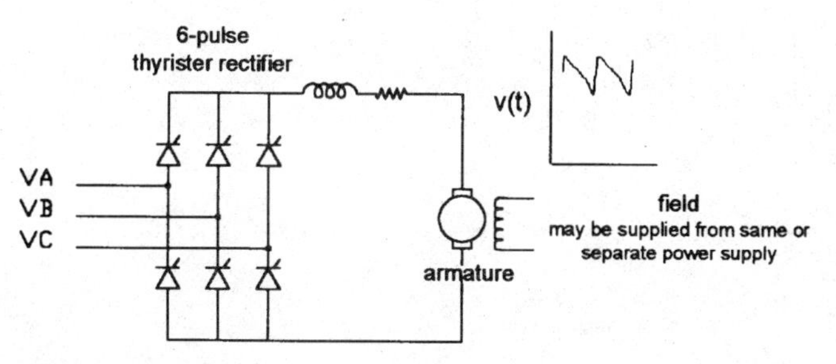

Figure 5.11 Six-pulse dc ASD.

pulse drive are the 5th and the 7th. They are also the most troublesome in terms of system response. A 12-pulse rectifier in this application can be expected to eliminate about 90 percent of the 5th and 7th harmonics, depending on system imbalances. The disadvantage of the 12-pulse drive is that there is more cost in electronics and another transformer is generally required.

5.8.2 ac drives

In ac drives, the rectifier output is inverted to produce a variable-frequency ac voltage for the motor. Inverters are classified as voltage source inverters (VSI) or current source inverters (CSI). A VSI requires a constant dc (i.e., low-ripple) voltage input to the inverter stage,. This is achieved with a capacitor or LC filter in the dc link. The CSI requires a constant current input; hence a series inductor is placed in the dc link.

AC drives generally use standard squirrel cage induction motors. These motors are rugged, relatively low in cost, and require little maintenance. Synchronous motors are used where precise speed control is critical.

A popular ac drive configuration uses a VSI employing PWM techniques to synthesize an ac waveform as a train of variable-width dc pulses (Fig. 5.12). The inverter uses either SCRs, gate

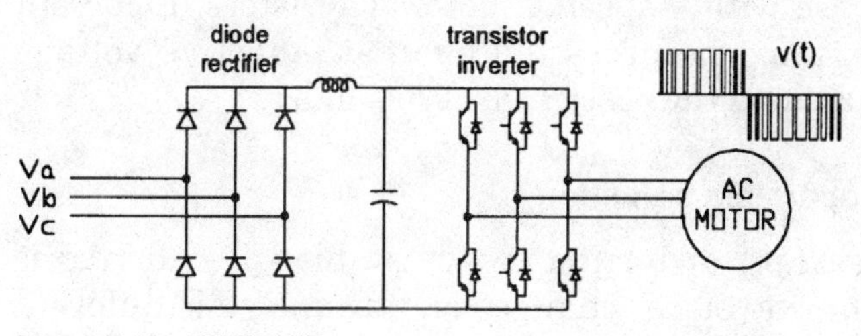

Figure 5.12 PWM ASD.

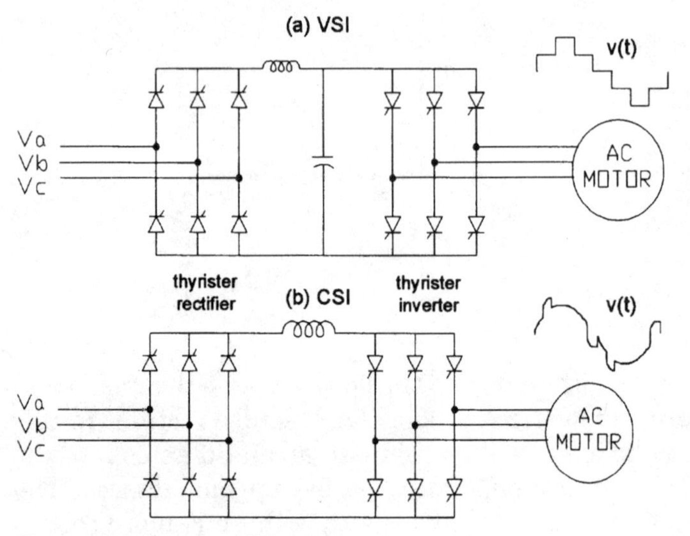

Figure 5.13 Large ac ASDs.

turnoff (GTO) thyristors, or power transistors for this purpose. Currently, the VSI PWM drive offers the best energy efficiency over widespeed range applications for drives up through at least 500 hp. Another advantage of PWM drives is that, unlike other types of drives, it is not necessary to vary rectifier output voltage to control motor speed. This allows the rectifier thyristors to be replaced with diodes, and the thyristor control circuitry to be eliminated.

Very-high-power drives employ SCR rectifiers and inverters. These may be six-pulse, as shown in Fig. 5.13, or like large dc drives, 12-pulse. VSI drives (Fig. 5.13*a*) are limited to applications that do not require rapid changes in speed. CSI drives (Fig. 5.13*b*) have good acceleration/deceleration characteristics, but require a motor with leading power factor (synchronous or induction with capacitors) or added control circuitry to commutate the inverter thyristors. In either case, the CSI drive must be designed for use with a specific motor. Thyristors in current source inverters must be protected against inductive voltage spikes, which increases the cost of this type of drive.

5.8.3 Impact of operating condition

The harmonic current distortion in adjustable-speed drives is not constant. The waveform changes significantly for different speed and torque values.

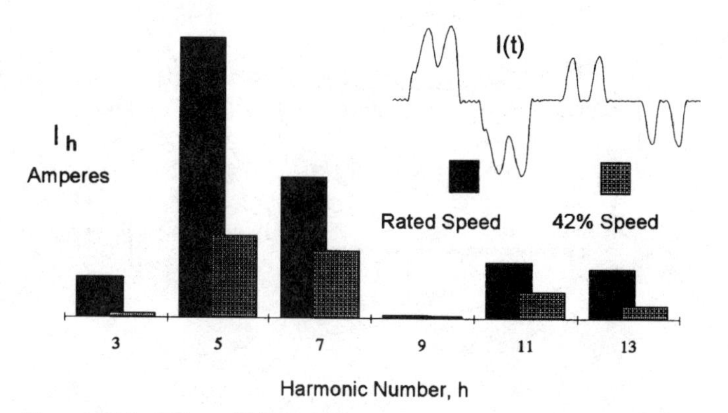

Figure 5.14 Effect of PWM ASD speed on ac current harmonics.

Figure 5.14 shows two operating conditions for a PWM adjustable-speed drive. While the waveform at 42 percent speed is much more distorted proportionately, the drive injects considerably higher magnitude harmonic currents at rated speed. The bar chart shows the amount of current injected. This will be the limiting design factor, not the highest THD. Engineers should be careful to understand the basis of data and measurements concerning these drives before making design decisions.

5.8.4 Effects of ac line chokes on harmonics

Inserting some additional reactance between an ASD and the system reduces the harmonic content of the ac line current. This is particularly effective for PWM drives. Figure 5.15 shows a plot of current distortion versus the ratio of drive kilovoltamperes to transformer kilovoltamperes for two different cases: with and without a 3 percent choke. The choke is rated on the base of the ASD. Representative waveforms for each end of the range are shown. The larger waveform is without the choke.

A very substantial improvement is possible by adding a choke, dropping the THD of the current from the 90 to 100 percent range down to the 30 to 40 percent range. The inductance slows the rate at which the capacitor on the dc bus can be charged and forces the drive to draw current over a longer time period. The net effect is a lower-magnitude current with much less harmonic content while still delivering the same energy.

Chokes also help eliminate nuisance drive tripping due to capacitor-switching transients.

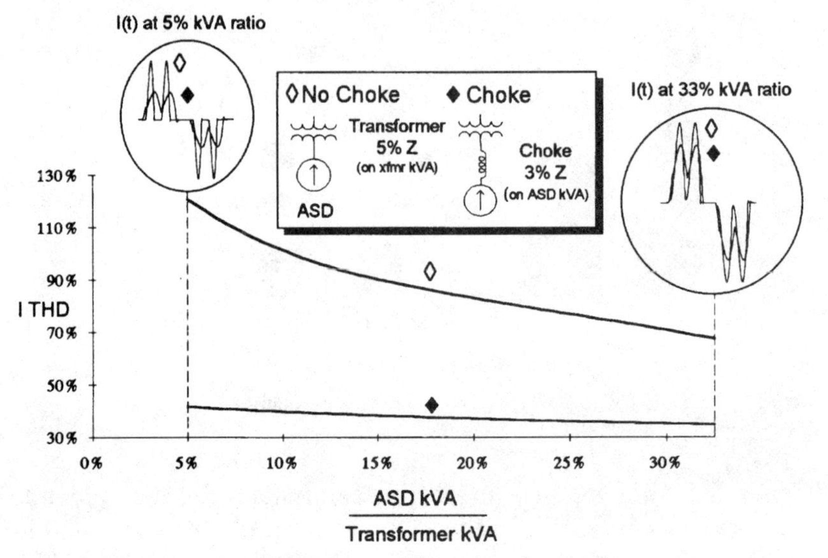

Figure 5.15 Effect of ac line chokes on ASD current harmonics.

5.9 Arcing Devices

This category includes arc furnaces, arc welders, and discharge-type lighting (fluorescent, sodium vapor, mercury vapor) with magnetic (rather than electronic) ballasts. As shown in Fig. 5.16, the arc is basically a voltage clamp in series with a reactance that limits current to a reasonable value.

The voltage-current characteristics of electric arcs are nonlinear. Following arc ignition, the voltage decreases as the arc current increases, limited only by the impedance of the power system. This gives the arc the appearance of having a negative resistance for a portion of its operating cycle. In fluorescent lighting applications, additional "ballasting" impedance is necessary to limit the current to within the capabilities of the fluorescent tube and stabilize the arc. Thus, this type of lighting has an external impedance element called a ballast. Magnetic

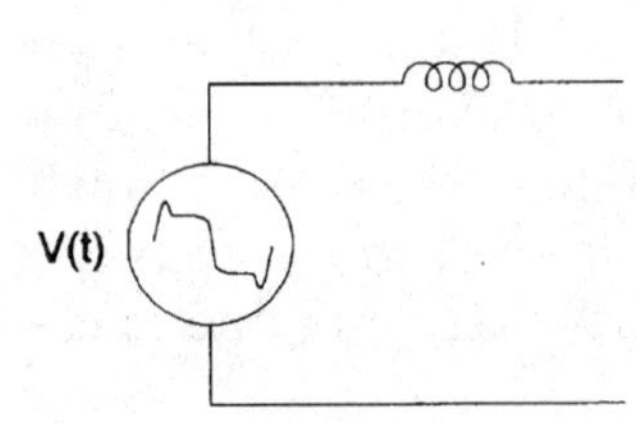

Figure 5.16 Equivalent circuit for an arcing device.

ballasts are usually rather benign sources of additional harmonics themselves; the main harmonic distortion comes from the behavior of the arc. However, some electronic ballasts, which may employ switch-mode power supplies for improved energy efficiency, may double or triple the normal harmonic output. Others have been specifically designed to minimize harmonics and may actually produce less harmonics than the normal magnetic ballast-lamp combination.

In electric arc furnace applications, the limiting impedance is primarily the furnace cable and leads with some contribution from the power system and furnace transformer. Currents in excess of 60,000 A are common.

The electric arc itself is actually best represented as a source of voltage harmonics. If a probe were to be placed directly across the arc, one would observe a somewhat trapezoidal waveform. Its magnitude is largely a function of the length of the arc. However, the impedance of ballasts or furnace leads acts as a buffer so that the supply voltage is only moderately distorted. The arcing load thus appears to be a relatively stable harmonic current source, which is adequate for most analysis. The exception occurs when the system is near resonance and a Thevenin equivalent model using the arc voltage waveform gives more realistic answers.

Figure 5.17 shows a measured fluorescent lamp current and harmonic spectrum. This lamp had a magnetic ballast. The harmonic content of this waveform is also similar to that of an arc furnace load and other arcing devices. Three-phase arcing devices can be arranged to cancel the triplen harmonics through the transformer connection. However, you cannot depend on this cancellation in three-phase arc furnaces because of the frequent imbalanced operation during the melting phase. During the refining stage when the arc is more constant, the cancellation is better. Fluorescent lighting in commercial buildings can be distributed among the phases in a nearly balanced manner to minimize the amount of triplen harmonic currents flowing onto the power supply system. Keep in mind that the common wye-wye supply transformers will not impede the flow of triplen harmonics regardless of how well balanced the phases are.

5.10 Saturable Devices

Equipment in this category includes transformers and other electromagnetic devices with a steel core, including motors.

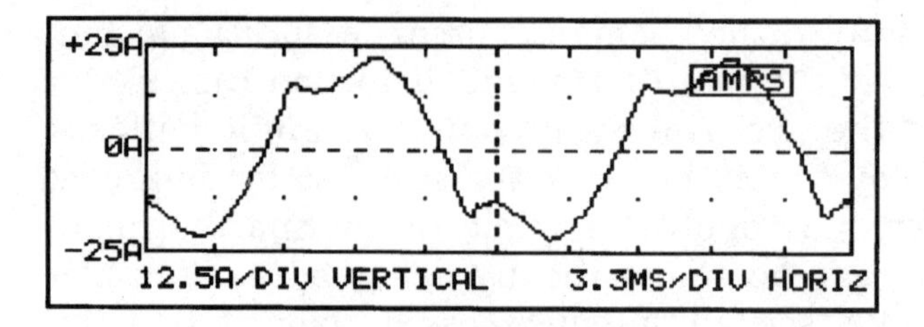

Harmonic	Percent	Phase (deg)
Fund	100.0	124
2	0.2	136
3	19.9	-144
5	7.4	62
7	3.2	-39
9	2.4	-171
11	1.8	111
13	0.8	17
15	0.4	-93
17	0.1	-164
19	0.2	-99
21	0.1	160

Figure 5.17 Fluorescent light current and harmonic spectrum.

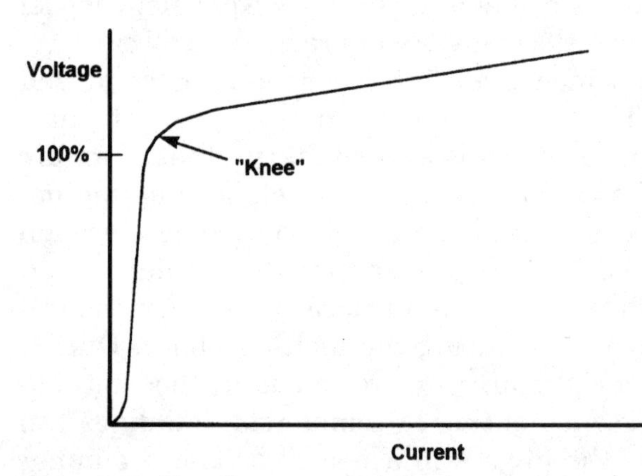

Figure 5.18 Transformer magnetizing characteristic.

Harmonics are generated due to the nonlinear magnetizing characteristics of the steel (Fig. 5.18).

Power transformers are designed to normally operate just below the "knee" point of the magnetizing saturation characteristic. The operating flux density of a transformer is selected based on a complicated optimization of steel cost, no-load losses, noise, and numerous other factors. Many electric utilities penalize transformer vendors by various amounts for no-load and load losses, and the vendor tries to meet the specification with a transformer that has the lowest evaluated cost. A high cost penalty on the no-load losses or noise generally results in more steel in the core and a higher saturation curve that yields lower harmonic currents.

Although transformer exciting current is rich in harmonics at normal operating voltage (Fig. 5.19), it is typically less than 1 percent of rated full-load current. Transformers are not as much of a concern as electronic power converters and arcing devices, which can produce harmonic currents of 20 percent of their rating or higher. However, their effect will be noticeable, particularly on utility distribution systems, which have hun-

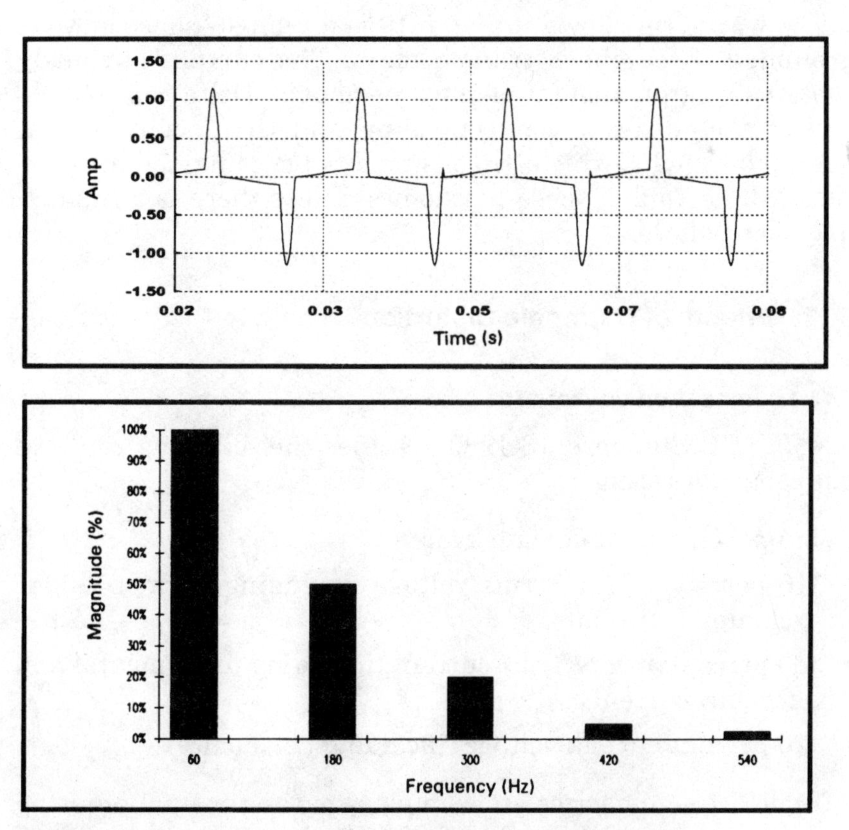

Figure 5.19 Transformer magnetizing current and harmonic spectrum.

dreds of transformers. It is common to notice a significant increase in triplen harmonic currents during the early morning hours when the load is low and the voltage rises. Transformer exciting current is more visible then because there is insufficient load to obscure it and the increased voltage causes more current to be produced. Harmonic voltage distortion from transformer overexcitation is generally apparent only under these light load conditions.

Some transformers are purposefully operated in the saturated region. One example is a triplen transformer used to generate 180 Hz for induction furnaces.

Motors also exhibit some distortion in the current when overexcited, although it is generally of little consequence. There are, however, some fractional-horsepower, single-phase motors that have a nearly triangular waveform with significant third-harmonic currents.

The waveform shown in Fig. 5.19 is for single-phase or wye-grounded three-phase transformers. The current obviously contains a large amount of third harmonic. Delta connections and ungrounded-wye connections prevent the flow of zero-sequence harmonic, which triplens tend to be. Thus, the line current will be void of these harmonics unless there is an imbalance somewhere.

5.11 Effects of Harmonic Distortion

5.11.1 Impact on capacitors

ANSI/IEEE Standard 18-1980 specifies the following continuous capacitor ratings:

- 135 percent of nameplate kvar
- 110 percent of rated rms voltage (including harmonics but excluding transients)
- 180 percent of rated rms current (including fundamental and harmonic current)
- 120 percent of peak voltage (including harmonics)

Table 5.1 summarizes an example capacitor evaluation using a computer spreadsheet that is designed to help evaluate the various capacitor duties against the standards.

TABLE 5.1 Example Capacitor Evaluation

Recommended practice for establishing capacitor capabilities when supplied by nonsinusoidal voltages (IEEE Standard 18-1980)

Capacitor bank data:

Bank rating:	1,200	kvar
Voltage rating:	13,800	V (LL)
Operating voltage:	13,800	V (LL)
Supplied compensation:	1,200	kvar
Fundamental current rating:	50.2	A
Fundamental frequency:	60	Hz
Capacitive reactance:	158,700	Ω

Harmonic distribution of bus voltage:

Harmonic no.	Frequency (Hz)	Volt. mag., V_h (% of fund.)	Volt. mag., V_h (V)	Line current, I_h (% of fund.)
1	60	100.00	7967.4	100.00
3	180	0.00	0.0	0.00
5	300	4.00	318.7	20.00
7	420	3.00	239.0	21.00
11	660	0.00	0.0	0.00
13	780	0.00	0.0	0.00
17	1020	0.00	0.0	0.00
19	1140	0.00	0.0	0.00
21	1260	0.00	0.0	0.00
23	1380	0.00	0.0	0.00
25	1500	0.00	0.0	0.00

Volt distortion (THD):	5.00 %
rms capacitor voltage:	7977.39 V
Capacitor current distortion:	29.00 %
rms capacitor current:	52.27 A

Capacitor bank limits:

	Calculated (%)	Limit (%)	Exceeds limit
Peak voltage	107.0	120	No
rms voltage	100.1	110	No
rms current	104.1	180	No
kvar	104.3	135	No

The fundamental full-load current for the 1200-kvar capacitor bank is determined from:

$$I_c = \frac{\text{kvar}_{3\phi}}{\sqrt{3} \times \text{kV}_{\phi\phi}} = \frac{1200}{\sqrt{3} \times 13.8} = 50.2 \text{ A} \tag{5.12}$$

The capacitor is subjected principally to two harmonics: the fifth and the seventh. The voltage distortion consists of 4 percent fifth and 3 percent seventh. This results in 20 percent fifth-harmonic current and 21 percent seventh-harmonic current. The resultant values all come out well below standard limits in this case, as shown in the box at the bottom of the table.

5.11.2 Impact on transformers

Transformers are designed to deliver the required power to the connected loads with minimum losses at fundamental frequency. Harmonic distortion of the current, in particular, as well as the voltage will contribute significantly to additional heating. To design a transformer to accommodate higher frequencies, designers make make different design choices such as using continuously transposed cable instead of solid conductor and putting in more cooling ducts. As a general rule, a transformer in which the current distortion exceeds 5 percent is a candidate for derating for harmonics.

There are three effects that result in increased transformer heating when the load current includes harmonic components:

1. *rms current.* If the transformer is sized only for the kVA requirements of the load, harmonic currents may result in the transformer rms current being higher than its capacity. The increased total rms current results in increased conductor losses.

2. *Eddy-current losses.* These are induced currents in a transformer caused by the magnetic fluxes. These induced currents flow in the windings, in the core, and in other conducting bodies subjected to the magnetic field of the transformer and cause additional heating. This component of the transformer losses increases with the square of the frequency of the current causing the eddy currents. Therefore, this becomes a very important component of transformer losses for harmonic heating.

TABLE 5.2 Simplified Example C57.110 Transformer Evaluation and K-Factor Computation

Site: Example Plant

Example Transformer

Harmonic distribution of transformer load current:

Harmonic	Current (%)	Frequency (Hz)	Current (pu)	I^2	$I^2 \times h^2$
1	100.000	60	1.000	1.000	1.000
3	1.600	180	0.016	0.000	0.002
5	26.100	300	0.261	0.068	1.703
7	5.000	420	0.050	0.003	0.123
9	0.300	540	0.003	0.000	0.001
11	8.900	660	0.089	0.008	0.958
13	3.100	780	0.031	0.001	0.162
15	0.200	900	0.002	0.000	0.001
17	4.800	1020	0.048	0.002	0.666
19	2.600	1140	0.026	0.001	0.244
21	0.100	1260	0.001	0.000	0.000
23	3.300	1380	0.033	0.001	0.576
25	2.100	1500	0.021	0.000	0.276
			Totals:	1.084	5.712
				K factor:	5.3

Standard derating (ANSI/IEEE C57.110-1986): 0.87 pu

Assumed eddy-current loss factor ($P_{EC\text{-}R}$) = 8%

3. *Core losses.* The increase in core losses in the presence of harmonics will be dependent on the effect of the harmonics on the applied voltage and the design of the transformer core. Increasing the voltage distortion may increase the eddy currents in the core laminations. The net impact that this will have depends on the thickness of the core laminations and the quality of the core steel. The increase in these losses due to harmonics is generally not as critical as the previous two.

Guidelines for transformer derating are detailed in ANSI/IEEE Standard C57.110. Table 5.2 illustrates a simplified approach. The common *K* factor used in the power quality field for transformer derating is also included in Table 5.2.[2]

The analysis represented in this table can be summarized as follows. The load loss, P_{LL}, can be considered to have two components: I^2R loss and eddy-current loss, P_{EC}:

$$P_{LL} = I^2R + P_{EC} \qquad (W) \tag{5.13}$$

The I^2R loss is directly proportional to the rms value of the current. However, the eddy current is proportional to the square of the current and frequency, which we can define by

$$P_{EC} = K_{EC} \times I^2 \times h^2 \tag{5.14}$$

where K_{EC} = proportionality constant. The per-unit full-load loss under harmonic current conditions is given by

$$P_{LL} = \sum I_h^2 + \left(\sum I_h^2 \times h^2\right) P_{EC\text{-}R} \tag{5.15}$$

where $P_{EC\text{-}R}$ = eddy-current loss factor under rated conditions.

The K factor[2] commonly found in power quality literature concerning transformer derating can be defined solely in terms of the harmonic currents as follows:

$$K = \frac{\sum (I_h^2 \times h^2)}{\sum I_h^2} \tag{5.16}$$

Then, in terms of the K factor, the rms of the distorted current is derived to be

$$\sqrt{\sum I_h^2} = \sqrt{\frac{1 + P_{EC\text{-}R}}{1 + K \times P_{EC\text{-}R}}} \quad \text{(pu)} \tag{5.17}$$

where $P_{EC\text{-}R}$ = eddy-current loss factor
h = harmonic number
I_h = harmonic current

Thus, the transformer derating can be estimated by knowing the per unit eddy-current loss factor. This factor can be determined by

1. Obtaining the factor from the transformer designer.
2. Using transformer test data and the procedure in ANSI/IEEE Standard C57.110.
3. Typical values based on transformer type and size (see Table 5.3).

TABLE 5.3 Typical Values of P_{EC-R}

Type	MVA	Voltage	$\%P_{EC-R}$
Dry	≤ 1		3–8
	≥ 1.5	5 kV HV	12–20
	≤ 1.5	15 kV HV	9–15
Oil-filled	≤ 2.5	480 V LV	1
	2.5 to 5	480 V LV	1–5
	>5	480 V LV	9–15

SOURCE: D. E. Rice, "Adjustable-Speed Drive and Power Rectifier Harmonics: Their Effects on Power System Components," in *Proceedings of the IEEE PCIC Conference,* No. PCIC-84-52.

Exceptions. There are often cases involving transformers that do not appear to have a harmonics problem from the above criteria, yet are running hot or failing due to what appears to be overload. One common case found with grounded-wye transformers is that the line currents contain about 8 percent third harmonic, which is relatively low, and the transformer is overheating at less than rated load. Why would this transformer pass the heat run test in the factory, and, perhaps, an overload test also, and fail to perform as expected in practice? Discounting mechanical cooling problems, chances are good that there is some conducting element in the magnetic field that is being affected by the harmonic fluxes. Three of several possibilities are:

1. Zero-sequence fluxes will "escape" the core on three-legged core designs (the most popular design for utility distribution substation transformers). This is illustrated in Fig. 5.20. The

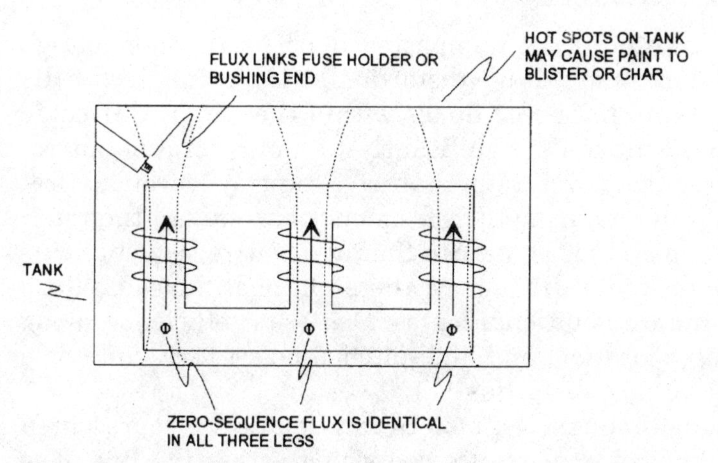

Figure 5.20 Zero-sequence flux in three-legged core transformers enters the tank, air, and oil space.

3d, 9th, 15th, etc., harmonics are predominantly zero-sequence. Therefore, if the winding connections are proper to allow zero-sequence current flow, these harmonic fluxes can cause additional heating in the tanks, core clamps, etc., that would not necessarily be found under balanced three-phase tests or single-phase tests. The 8 percent line current mentioned above translates to a neutral current third-harmonic current of 24 percent of the phase current. This could add considerably to the leakage flux in the tank and in the oil and air space. Two symptoms are charred or bubbled paint on the tank and evidence of heating on the end of a bayonet fuse tube (without blowing the fuse) or bushing end.

2. dc offsets in the current can also cause flux to "escape" the confines of the core. The core will become slightly saturated on, for example, the positive half-cycle while remaining normal for the negative half-cycle. There are a number of electronic power converters that produce current waveforms that are nonsymmetrical either by accident or by design. This can result in a small dc offset on the load side of the transformer (it can't be measured from the source side). Only a small amount of dc offset is required to cause problems with most power transformers.

3. There may be a clamping structure, bushing end, or some other conducting element too close to the magnetic field. It may be sufficiently small in size that there is no notable effect in stray losses at fundamental frequency, but may produce a hot spot when subjected to harmonic fluxes.

5.11.3 Impact on motors

Motors can be significantly impacted by the harmonic voltage distortion. Harmonic voltage distortion at the motor terminals is translated into harmonic fluxes within the motor. Harmonic fluxes do not contribute significantly to motor torque, but rotate at a frequency different than the rotor synchronous frequency, basically inducing high-frequency currents in the rotor. The effect on motors is similar to that of negative sequence currents at fundamental frequency: the additional fluxes do little more than induce additional losses. Decreased efficiency, along with heating, vibration, and high-pitched noises, are symptoms of harmonic voltage distortion.

At harmonic frequencies, motors can usually be represented by the blocked rotor reactance connected across the line. The lower-order harmonic voltage components, for which the magnitudes are larger and the apparent motor impedance lower, are usually the most important for motors.

There is usually no need to derate motors if the voltage distortion remains within IEEE Standard 519-1992 limits of 5 percent THD and 3 percent for any individual harmonic. Excessive heating problems begin when the voltage distortion reaches 8 to 10 percent and higher. Such distortion should be corrected for long motor life.

Motors appear to be in parallel with the power system impedance with respect to the harmonic current flow and generally shift the system resonance higher by causing the net inductance to decrease. Whether this is detrimental to the system depends on the location of the system resonance prior to energizing the motor. Motors also may contribute some to the damping of some of the harmonic components depending on the X/R ratio of the blocked rotor circuit. In systems with many smaller-sized motors, which have a low X/R ratio, this could help attenuate harmonic resonance. However, you cannot depend on this with large motors.

5.12 System Response Characteristics

In power systems, the response of the system is equally as important as the sources of harmonics. In fact, power systems are quite tolerant of the currents injected by harmonic-producing loads unless there is some adverse interaction with the impedance of the system. Identifying the sources is only half the job of harmonic analysis. The response of the power system at each harmonic frequency determines the true impact of the nonlinear load on harmonic voltage distortion.

5.12.1 System impedance

At the fundamental frequency, power systems are primarily inductive, and the equivalent impedance is sometimes called simply the *short-circuit reactance.* Capacitive effects are frequently neglected on utility distribution systems and industrial power systems. One of most frequently used quantities in the analysis of harmonics on power systems is the short-circuit impedance to the point on a network at which a capacitor is located. If not directly available, it can be computed from short-circuit study results that give either the short-circuit megavoltampere (MVA) or the short-circuit current as follows:

$$Z_{SC} = R_{SC} + jX_{SC}$$

$$= \frac{kV^2}{MVA_{SC}} = \frac{I_{SC}}{\sqrt{3}\ kV} \tag{5.18}$$

where Z_{SC} = short-circuit impedance
R_{SC} = short-circuit resistance
X_{SC} = short-circuit reactance
kV = phase-to-phase voltage, kV
MVA_{SC} = three-phase short-circuit MVA
I_{SC} = short-circuit current, A

Z_{SC} is a phasor quantity, consisting of both resistance and reactance. However, if the short-circuit data contain no phase information, one is usually constrained to assuming that the impedance is purely reactive. This is a reasonably good assumption for industrial power systems for buses close to the mains and for most utility systems. When this is not the case, an effort should be made to determine a more realistic resistance value because that will affect the results once capacitors are considered.

The inductive reactance portion of the impedance changes linearly with frequency. One common error made by novices in harmonic analysis is to forget to adjust the reactance for frequency. The reactance at the *h*th harmonic is determined from the fundamental-impedance reactance, X_1, by

$$X_h = hX_1 \tag{5.19}$$

In most power systems, one can generally assume that the resistance does not change significantly when studying the effects of harmonics less than the ninth. For lines and cables, the resistance varies approximately by the square root of the frequency once skin effect becomes significant in the conductor at a higher frequency. The exception to this rule is with some transformers. Because of stray eddy current losses, the apparent resistance of larger transformers may vary almost proportionately with the frequency. This can have a very beneficial effect on damping of resonance as we shall see below. In smaller transformers, less than 100 kVA, the resistance of the winding is often so large relative to the other impedances that it swamps out the stray eddy current effects and there is little change in the total apparent resistance until the frequency reaches about 500 Hz. Of course, these smaller transformers may have an X/R ratio of 1.0 to 2.0 at fundamental frequency while large substation transformers might typically be 20 to 30. Therefore, if the bus that is being studied is dominated by transformer impedance rather than line impedance, the system impedance model should be considered more carefully. Neglecting the resistance will generally give a conservatively high prediction of the harmonic distortion.

At utilization voltages, such as industrial power systems, the equivalent system reactance is often dominated by the service transformer impedance. A good approximation for X_{SC} may be based on the impedance of the transformer only:

$$X_{SC} \approx X_{tx} \tag{5.20}$$

While not precise, this is generally at least 90 percent of the total impedance and is commonly more. This is usually sufficient to evaluate whether or not there will be a significant harmonic resonance problem. Transformer impedance in ohms can be determined from the percent impedance, Z_{tx}, found on the nameplate by

$$X_{tx} = \left(\frac{\text{kV}^2_{\phi\phi}}{\text{MVA}_{3\phi}}\right) \times Z_{tx}\,(\%) \tag{5.21}$$

This assumes that the impedance is predominantly reactive. For example, for a 1500-kVA, 6 percent transformer, the equivalent impedance on the 480-V side is

$$X_{tx} = \left(\frac{\text{kV}^2_{\phi\phi}}{\text{MVA}_{3\phi}}\right) \times Z_{tx}\,(\%) = \left(\frac{0.480^2}{1.5}\right) \times 0.06 = 0.0092\ \Omega \tag{5.22}$$

A plot of impedance vs. frequency for an inductive system (no capacitors installed) would look like Fig. 5.21.

Real power systems are not quite as well behaved. This simple model neglects capacitance, which cannot be done for harmonic analysis.

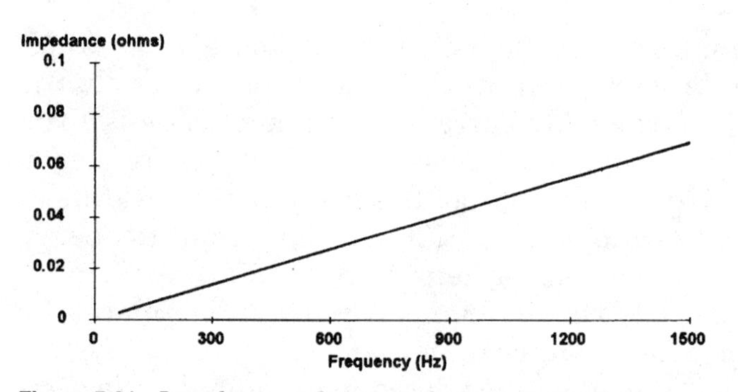

Figure 5.21 Impedance vs. frequency for inductive system.

5.12.2 Capacitor impedance

Shunt capacitors, either at the customer location for power factor correction, or on the utility distribution system, dramatically alter the system impedance variation with frequency. Capacitors do not create harmonics, but severe harmonic distortion can sometimes be attributed to their presence. While the reactance of inductive components increases proportionately to frequency, capacitive reactance, X_c, decreases proportionately:

$$X_c = \frac{1}{2\pi \times f \times C} \tag{5.23}$$

where C is the capacitance in farads. This quantity is seldom readily available for power capacitors, which are rated in terms of kvar or Mvar at a given voltage. The equivalent line-to-neutral capacitive reactance at fundamental frequency for a capacitor bank can be determined by

$$X_c = \frac{\text{kV}^2}{\text{Mvar}} = \frac{\text{kV}^2(1000)}{\text{kvar}} \tag{5.24}$$

For three-phase banks, use phase-to-phase voltage and the three-phase reactive power rating. For single-phase units, use the can voltage rating and the reactive power rating. For example, for a three-phase, 1200-kvar, 13.8-kV capacitor bank, the positive-sequence reactance in ohms would be

$$X_c = \frac{\text{kV}^2}{\text{Mvar}} = \frac{13.8^2}{1.2} = 158.7\ \Omega \tag{5.25}$$

5.12.3 Parallel resonance

All circuits containing both capacitances and inductances have one or more natural frequencies. When one of those frequencies lines up with a frequency that is being produced on the power system, resonance can develop in which the voltages and current at that frequency continue to persist at very high values. This is the root of most problems with harmonic distortion on power systems.

At harmonic frequencies, from the perspective of harmonic sources, shunt capacitors appear to be in parallel with the equivalent system inductance as shown in the equivalent circuit in Fig. 5.22*a* and *b*. At frequencies other than fundamental, the power system generation appears to be a short circuit. It is assumed that it is a voltage source of fundamental frequency only, which is generally a good assumption. At the frequency where X_c and the total system reactance are equal, the apparent impedance of the parallel combination of inductance and capacitance as seen by the source of harmonic currents becomes very large. This results in the typical parallel resonance condition. The effect of varying capacitor size on the impedance seen from the harmonic source is shown in Fig. 5.22*c* and

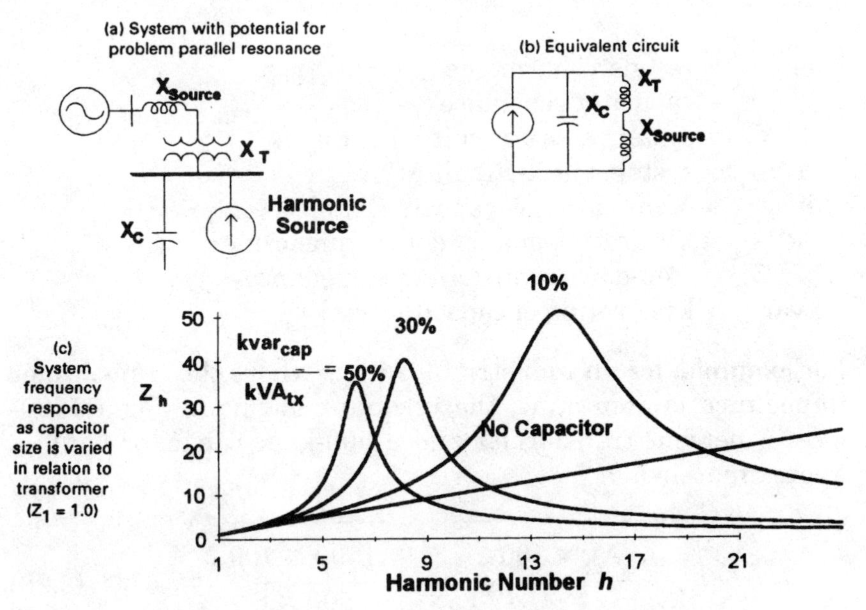

Figure 5.22 Effect of capacitor size on parallel resonant frequency.

compared with the case in which there is no capacitor. It is apparent that if one of the peaks lines up with a common harmonic current produced by the load, there will be a much higher voltage drop across the apparent impedance than without capacitors.

The resonant frequency for a particular inductance-capacitance combination can be computed from a variety of different formulae depending on what data are available. The basic resonant frequency equation is

$$f_r = \frac{1}{2\pi\sqrt{LC}} \tag{5.26}$$

Power systems analysts typically do not have L and C readily available and prefer to use other forms of this relationship. They commonly compute the resonant harmonic, h_r, based on fundamental frequency impedances and ratings using one of the following:

$$h_r = \sqrt{\frac{X_c}{X_{SC}}} = \sqrt{\frac{\text{MVA}_{SC}}{\text{Mvar}_{cap}}} \approx \sqrt{\frac{\text{kVA}_{tx} \times 100}{\text{kvar}_{cap} \times Z_{tx}\,(\%)}} \tag{5.27}$$

where h_r = resonant harmonic
X_c = capacitor reactance
X_{SC} = system short-circuit reactance
MVA_{SC} = system short-circuit MVA
Mvar_{cap} = Mvar rating of capacitor bank
kVA_{tx} = kVA rating of step-down transformer
Z_{tx} = step-down transformer impedance
kvar_{cap} = kvar rating of capacitor bank

For example, for an industrial load bus where the transformer impedance is dominant, the resonant harmonic for a 1500-kVA, 6 percent transformer and a 500-kvar capacitor bank is approximately

$$h_r \approx \sqrt{\frac{\text{kVA}_{tx} \times 100}{\text{kvar}_{cap} \times Z_{tx}\,(\%)}} = \sqrt{\frac{1500 \times 100}{500 \times 6}} = 7.07 \tag{5.28}$$

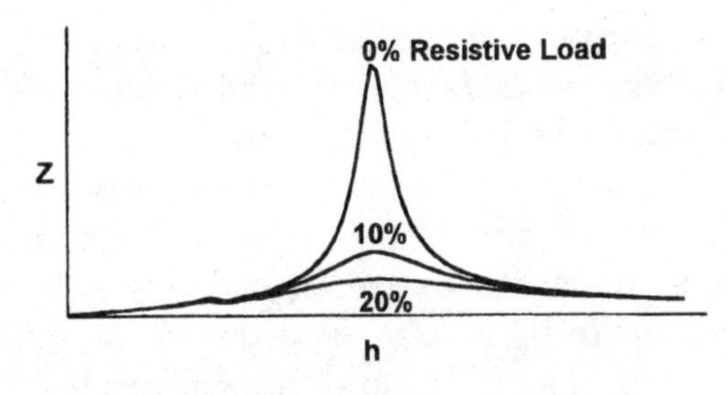

Figure 5.23 Effect of resistance loads on parallel resonance.

5.12.4 Effects of resistance and resistive load

Determining that the resonant harmonic aligns with a common harmonic source is not always cause for alarm. The damping provided by resistance in the system is often sufficient to prevent catastrophic voltages and currents. Figure 5.23 shows the parallel resonant circuit impedance characteristic for various amounts of resistive load in parallel with the capacitance. As little as 10 percent resistive loading can have a significant beneficial impact on peak impedance. Likewise, if there is a significant length of lines or cables between the capacitor bus and the nearest upline transformer, the resonance will be suppressed. Lines and cables can add a significant amount of the resistance to the equivalent circuit.

Loads and line resistances are why we seldom see catastrophic harmonic problems from capacitors on utility distribution feeders. That is not to say that there will not be any harmonic problems due to resonance, but that the problems will generally not cause physical damage to the electrical system components. The most troublesome resonant conditions occur when capacitors are installed on substations buses, either utility substations or in industrial facilities. In these cases, where the transformer dominates the system impedance and has a high X/R ratio, the relative resistance is low and the corresponding parallel resonant impedance peak is very sharp and high. This is a common cause of capacitor failure, transformer failure, or the failure of load equipment.

While utility distribution engineers may be able to place feeder banks with little concern about resonance, studies should always be performed for industrial capacitor applications and for utility substation applications. Utility engineers have told us that about 20 percent of industrial installations for which no studies are performed have major operating disruptions or equipment failure due to resonance. In fact, selecting capacitor sizes from manufacturers' tables to correct the power factor based on average monthly billing data tends to result in a combination that tunes the system near the fifth harmonic. This is one of the worst harmonics to which to be tuned because it is frequently the largest component in three-phase systems.

It is a misconception that resistive loads damp harmonics because in the absence of resonance, loads of any kind will have little impact on the harmonic currents and resulting voltage distortion. Most of the current will flow back in to the power source. However, it is very appropriate to say that resistive loads will damp *resonance,* which will lead to a significant reduction in the harmonic distortion.

Motor loads are primarily inductive and provide little damping. In fact, they may increase distortion by shifting the system resonant frequency closer to a significant harmonic. Small, fractional-horsepower motors may contribute significantly to damping because their apparent X/R ratio is lower than large three-phase motors.

5.13 Principles for Controlling Harmonics

This section describes some of the basic principles for controlling harmonics.

Fundamentally, harmonics become a problem if:

1. The source of harmonic currents is too great.
2. The path in which the currents flow is too long (electrically), resulting in either high voltage distortion or telephone interference.
3. The response of the system accentuates one or more harmonics.

When a problem occurs, the basic options for controlling harmonics are:

1. Reduce the harmonic currents produced by the load.
2. Add filters to either siphon the harmonic currents off the system, block the currents from entering the system, or supply the harmonic currents locally.

3. Alter the frequency response of the system by filters, inductors, and capacitors.

5.13.1 Reducing harmonic currents in loads

There is often little that can be done with existing load equipment to significantly reduce the amount of harmonic it is producing unless it is being misoperated. While an overexcited transformer can be brought back into normal operation by lowering the applied voltage, arcing devices and most electronic power converters are locked into their designed characteristic.

PWM drives that charge the dc bus capacitor directly from the line without any intentional impedance are one exception to this. Adding a line reactor in series will significantly reduce harmonics, as well as provide transient protection benefits.

Transformer connections can be employed to reduce harmonic in three-phase systems. Phase-shifting half of the six-pulse power converters in a plant load by 30 degrees can approximate the benefits of 12-pulse loads by dramatically reducing the fifth and seventh harmonics. Delta-connected transformers can block the flow of zero-sequence harmonics (typically triplens) from the line. Zigzag and grounding transformers can shunt the triplens off the line.

Purchasing specifications can go a long way toward preventing harmonic problems by penalizing bids from vendors with high harmonic content. This is particularly important in such loads as high-efficiency lighting.

5.13.2 Filtering

The shunt filter works by short-circuiting the harmonic currents as close to the source of distortion as practical. This keeps the currents out of the supply system. This is the most common type of filtering applied because of economics and because it also tends to smooth the load voltage as well as remove the harmonic current.

Another approach is to apply a series filter that blocks the harmonic currents. This is a parallel-tuned circuit that offers a high impedance to the harmonic current. It is not often used because it is difficult to insulate and the load voltage is very distorted. One common application is in the neutral of a grounded-wye capacitor to block the flow of triplen harmonics while still retaining a good ground at fundamental frequency.

Active filters work by electronically supplying the harmonic component of the current into a nonlinear load.

More information on filtering is given in Sec. 5.15.

5.13.3 Modifying the system frequency response

Adverse system responses to harmonics can be modified by a number of methods:

1. Adding a shunt filter. Not only does this shunt a troublesome hamonic current off the system, but it also completely changes the system response, most often, but not always, for the better.
2. Adding a reactor to detune the system. Harmful resonances are generally between the system inductance and shunt power factor correction capacitors. The reactor must be added between the capacitor and the system. One method is to simply put a reactor in series with the capacitor to move the system resonance without actually tuning the capacitor to create a filter.
3. Changing the size of the capacitor. This is often one of the least expensive options for both utilities and industrial customers.
4. Moving a capacitor to a point on the system with a different short-circuit impedance or higher losses. This is also an option for utilities when adding a bank causes telephone interference—moving the bank to another branch of the feeder may very well resolve the problem. This is frequently not an option for industrial users because the capacitor cannot be moved far enough to make a difference.
5. Removing the capacitor and simply accepting the higher losses, lower voltage, and power factor penalty. If technically feasible, this is occasionally the best economic choice.

5.13.4 On utility distribution feeders

The X/R ratio of the utility distribution feeder is generally low. Therefore, the accentuation of harmonics by resonance with feeder banks is usually mild. However, it may be very noticeable when a capacitor bank is energized and can still cause equipment malfunctions. Utility distribution engineers can usually place feeder banks where they wish without excessive concern about harmonics. When problems do occur, the usual solution is to move the bank or change the capacitor size.

Many harmonic problems due to adding feeder banks are due to increasing the triplen harmonics in the neutral circuit of the feeder. To change the flow of zero-sequence harmonic currents, changes are made to the neutral connection of wye-connected banks. To block the flow, the neutral is allowed to float.

Sometimes it is advantageous to put a reactor in the neutral to turn the bank into a tuned resonant shunt for a zero-sequence harmonic.

Many times, harmonic problems on distribution feeders exist only at light load. The voltage rises, causing the transformers to produce more harmonics. There is less load to damp out resonance. Switching the capacitors off at this time frequently solves the problem.

Should harmonic currents from widely dispersed sources require filtering on distribution feeders, the general idea is to distribute a few filters out on the feeder. This shortens the average path for the harmonic currents, reducing the opportunity for telephone interference and reducing the harmonic voltage drop in the lines. This keeps the voltage distortion out on the feeder to a minimum. With the ends of the feeder "nailed down" by filters with respect to the voltage distortion, it is more difficult for the voltage distortion to rise above limits elsewhere.

Harmonic studies should be performed on any large capacitor banks installed in distribution substations. One cannot count on system losses to damp out resonance at this point on the system. Placing a filter at the substation bank will not necessarily resolve the problems on the feeder unless the problem was based on resonance with a substation capacitor bank.

5.13.5 In end-user facilities

First, determine if a different size of capacitor can be used. Sometimes, there are so many capacitors switched with loads that it is impossible to control the value of capacitance. However, with switched capacitors and automatic power factor controllers, it may be possible to select a control scheme that avoids the configuration that causes problems.

Filter installation on end-user systems is more attractive practically and economically than on utility distribution systems. The criteria for filter installation are more easily met, and filtering equipment is more readily available on the market.

Industrial users should also investigate means of reducing harmonics by using different transformer connections. There are zigzag transformers and other devices available to remove triplen harmonics from three-phase circuits in office buildings.

Studies should be performed on all capacitors installed on industrial systems. The systems are generally so short that there are insufficient line losses to damp resonance if it exists. Some plants are exceptions to this because the capacitors are installed near the loads and there is sufficient resistance in lines to prevent problems. Also, some loads contribute significantly to damping. However, one should not count on this being the case until a study is performed. If one is installing capacitors for the first time, keep in mind that resonance problems are usually less severe when the capacitors are moved out onto the plant floor to motors and motor control centers. This also has the benefit of reducing the losses in the system over simply placing the capacitor on the main bus.

5.14 Locating Sources of Harmonics

On radial utility distribution feeders and industrial plant power systems, the main tendency is for the harmonic currents to flow from the harmonic-producing load to the power system source. This is illustrated in Fig. 5.24. The impedance of the power system is normally the lowest impedance seen by the harmonic currents. Thus, the bulk of the current flows into the source.

You can exploit this general tendency to locate sources of harmonics. Using a power quality monitor capable of reporting the harmonic content of the current, simply measure the harmonic

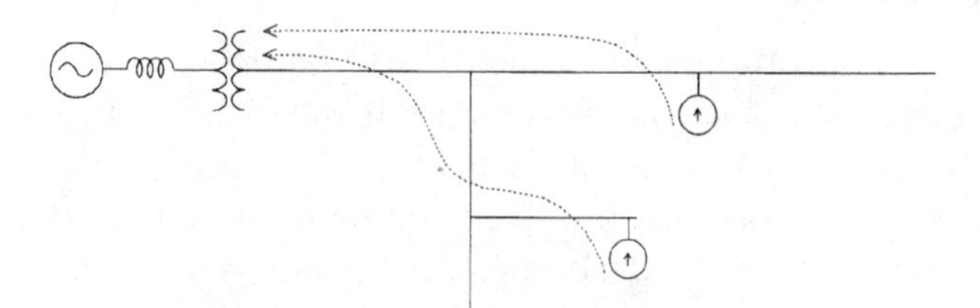

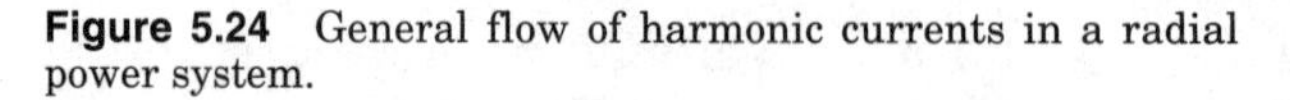

Figure 5.24 General flow of harmonic currents in a radial power system.

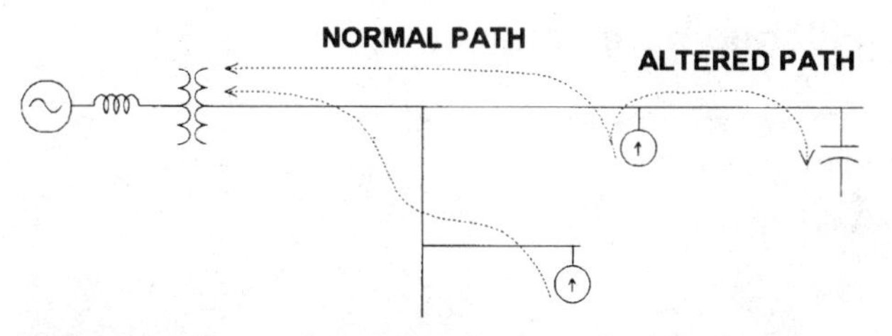

Figure 5.25 Power factor capacitors can alter the direction of flow of one of the harmonic components of the current.

currents in each branch, starting at the beginning of the circuit, and trace the harmonics to the source.

Power factor correction capacitors can alter this flow pattern for at least one of the harmonics. For example, adding a capacitor to the previous circuit as shown in Fig. 5.25 may draw a large amount of harmonic current into that portion of the circuit. If you are using the procedure described above, you may be tempted to follow the altered path with leads to a capacitor bank instead of the actual source of the harmonics. Thus, it is generally necessary to temporarily disconnect all capacitors to reliably locate the sources of harmonics.

It is usually very easy to differentiate harmonic currents due to actual sources from harmonic currents that are strictly due to resonance involving a capacitor bank. The resonance currents have one dominant harmonic riding on top of the fundamental sine wave. If you study the shapes of the current waveforms of harmonic sources presented earlier in this chapter, there are none that produce a single harmonic frequency in addition to the fundamental. They have somewhat arbitrary waveshapes depending on the distorting phenomena, but they contain several harmonics in significant quantities. A single, large, significant harmonic nearly always signifies resonance.

This fact can be exploited to determine if harmonic resonance problems are likely to exist in a system with capacitors. Simply measure the current in the capacitors. If it contains a very large amount of one harmonic other than the fundamental, it is likely that the capacitor is participating in a resonant circuit within the power system. Always check the capacitor currents first in any installations where harmonic problems are suspected.

5.15 Devices for Filtering Harmonic Distortion

There are two general classes of filters:

1. Passive filters
2. Active filters

We will describe the salient features of each class in the following two subsections.

5.15.1 Passive filters

Passive filters are made of inductance, capacitance, and resistance elements. They are relatively inexpensive compared with other means for eliminating harmonic distortion, but they have the disadvantage of potential adverse interactions with the power system. They are employed either to shunt the harmonic currents off the line or to block their flow between parts of the system by tuning the elements to create a resonance at a selected harmonic frequency. Figure 5.26 shows several types of common filter arrangements.

The most common type of passive filter is the single-tuned "notch" filter. This is the most economical type and is frequently sufficient for the application. An example of a common 480-V filter arrangement is illustrated in Fig. 5.27. The notch filter is series-tuned to present a low impedance to a particular harmonic current. It is connected in shunt with the power system. Thus, harmonic currents are diverted from their normal flow path on the line into the filter. Notch filters can provide power factor correction in addition to harmonic suppression.

The figure shows a common delta-connected low-voltage capacitor bank converted into a filter by adding an inductance in series. In this case, the notch harmonic, h_{notch}, is related to the fundamental frequency reactances by

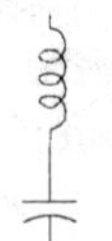

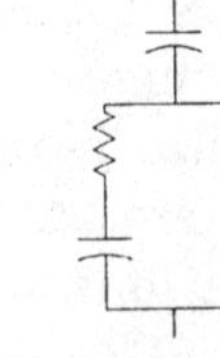

Figure 5.26 Common passive filter configuration.

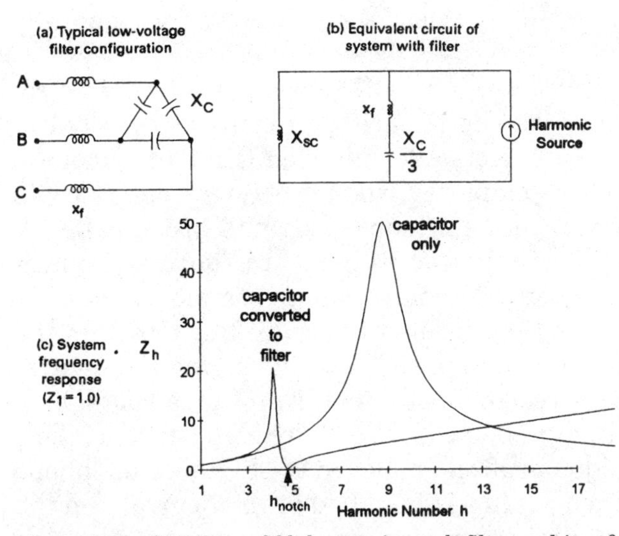

Figure 5.27 Creating a fifth-harmonic notch filter and its effect on system response.

$$h_{\text{notch}} = \sqrt{\frac{X_c}{3X_f}} \tag{5.29}$$

Note that X_c in this case is the reactance of one leg of the delta rather than the equivalent line-to-neutral capacitive reactance. If we were to use phase-to-phase voltage and three-phase kvar to compute X_c, as previously described, we would not divide by 3.

One important side effect of adding a filter is that it creates a sharp parallel resonance point at a frequency below the notch frequency (Fig. 5.27*c*). This resonant frequency must be safely away from any significant harmonic. Filters are commonly tuned slightly lower than the harmonic to be filtered to provide a margin of safety in case there is some change in system parameters. If they were tuned exactly to the harmonic, changes in either capacitance or inductance with temperature or failure might shift the parallel resonance higher into the harmonic. This could present a situation worse than without a filter because the resonance is generally very sharp.

For this reason, filters are added to the system starting with the lowest problem harmonic. For example, installing a seventh-harmonic filter usually requires that a fifth-harmonic filter also be installed. The new parallel resonance with a seventh filter only would have been very near the fifth, which is generally disastrous.

The filter configuration of Fig. 5.27*a* does not admit zero-sequence currents because the capacitor is connected in delta. This makes it largely ineffective for filtering triplen harmonics. Other solutions must be employed when it becomes necessary to control zero-sequence third-harmonic currents because 480-V capacitors are usually configured in delta. In contrast, capacitors on utility distribution systems are more commonly connected in wye. This gives the option of providing a path for the zero-sequence triplen harmonics simply by changing the neutral connection. Placing a reactor in the neutral of a capacitor is a common way to force the bank to filter only zero-sequence harmonics. This technique is often employed to eliminate telephone interference. A tapped reactor is installed in the neutral and the tap adjusted to minimize the telephone interference, depending on which harmonic is causing the problem.

Passive filters should always be placed on a bus where X_{SC} can be expected to remain constant. While the notch frequency will remain fixed, the parallel resonance will move with system impedance. For example, the parallel resonant frequency for running with standby generation by itself is likely to be much lower than when interconnected with the utility. Thus, filters are often removed for standby operation.

Also, filters must be designed with the capacity of the bus in mind. The temptation is to size the current-carrying capability based solely on the load that is producing the harmonic. However, a small amount of background voltage distortion on a very strong bus may impose excessive duty on the filter.

5.15.2 Active filters

Active filters are relatively new types of devices for eliminating harmonics. They are based on sophisticated power electronics and are much more expensive than passive filters. However, they have the distinct advantage that they do not resonate with the system. They can be used in very difficult circumstances where passive filters cannot operate successfully because of where the parallel resonance lies. They can also address more than one harmonic at a time and combat other power quality problems such as flicker. They are particularly useful for large, distorting loads fed from relatively weak points on the power system.

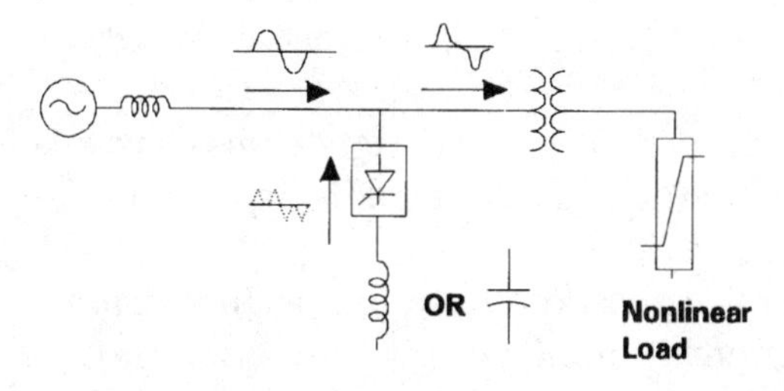

Figure 5.28 Application of an active filter at a load.

The basic idea is to replace the portion of the sine wave that is missing in the current in a nonlinear load. Figure 5.28 illustrates the concept. An electronic control monitors the line voltage and/or current, switching the power electronics very precisely to track the load current or voltage and force it to be sinusoidal. As shown, there are two fundamental approaches: one that uses an inductor to store up current to be injected into the system at the appropriate instant and one that uses a capacitor. Therefore, while the load current is distorted to the extent demanded by the nonlinear load, the current seen by the system is much more sinusoidal.

Active filters can typically correct for power factor as well as harmonics.

5.16 Harmonic Study Procedure

The following is the ideal procedure for performing a power systems harmonics study:

1. Determine the objectives of the study. This is important to keep the investigation on track. For example, the objective might be to identify what is causing an existing problem and solve it. Another objective might be to determine if a new plant expansion containing equipment like adjustable-speed drives and capacitors is likely to have problems.
2. Make a premeasurement computer simulation based on the best information available. Measurements are expensive in terms of time, equipment, and possible disruption to plant operations. It is generally economical to have a good idea what to look for and where to look before entering the facility.
3. Make measurements of the existing harmonic conditions, characterizing sources of harmonic and system bus distortion.
4. Calibrate the computer model using the measurements.

5. Study the new circuit condition or existing problem, whatever the case may be.
6. Develop solutions (filter, etc.) and investigate possible adverse system interactions. Also, check the sensitivity of the results to important variables.
7. After the installation of proposed solutions, perform monitoring to verify the correct operation of the system.

This procedure presumes access to computer analysis tools and adequate monitoring equipment.

Admittedly, it is not always possible to perform each of these steps to the ideal extent desired. The most often omitted steps are one or both measurement steps due to the cost of engineering time, travel, and equipment charges. An experienced analyst may be able to solve a problem without measurements, but it is strongly recommended that the initial measurements be made if at all possible because there are many unpleasant surprises lurking in the shadows of harmonics analysis.

5.17 Symmetrical Components

Power engineers have traditionally used symmetrical components to help them understand three-phase system behavior. The three-phase system is transformed into three single-phase systems that are much simpler to analyze. The method of symmetrical components can be employed for analysis of the system's response to harmonic currents provided care is taken not to violate the fundamental assumptions of the method.

The method allows any unbalanced set of phase currents (or voltages) to be transformed into three balanced sets. The *positive-sequence* set contains three sinusoids displaced 120 degrees from each other, with the normal A-B-C phase rotation. The sinusoids of the *negative-sequence* set are also displaced 120 degrees, but have opposite phase rotation (A-C-B). The sinusoids of the *zero sequence* are in phase with each other.

In *perfectly balanced* systems:

- Harmonics of order $h = 1, 7, 13, \ldots$ are purely positive-sequence.
- Harmonics of order $h = 5, 11, 17, \ldots$ are purely negative-sequence.
- Triplens ($h = 3, 9, 15, \ldots$) are purely zero-sequence.

In cases where the system is balanced, the terms “triplen” and “zero-sequence” are synonymous, but only when the system is balanced. When this condition is violated, any of the harmonics may be partially made up of any of the sequences.

The response of the system to positive-sequence harmonics is straightforward. It greatly simplifies matters if one only has to analyze the positive-sequence network, as both utility and industrial power engineers are accustomed to doing in their load flow and voltage drop analysis. Fortunately, there is a rule for many three-phase industrial-class loads that permits this. It may be simply stated:

> When there is a delta winding in a transformer anywhere in series with the harmonic source and the power system, only the positive-sequence circuit need be represented to determine the system response. It is impossible for zero-sequence harmonics to be present; they are blocked.

Figure 5.29 illustrates this principle, showing what models apply to different applications.

Both the positive- and negative-sequence networks generally have the same response to harmonics. The same circuit model may be used for either. If triplen harmonics do show up in measurements (they will for unbalanced sources), they will not be zero-sequence and can be analyzed with the same model.

The symmetrical component technique fails to yield an advantage when analyzing four-wire utility distribution feeders with numerous single-phase loads. Both the positive- and zero-sequence networks come into play. It is generally impractical to consider analyzing the system manually, and most computer

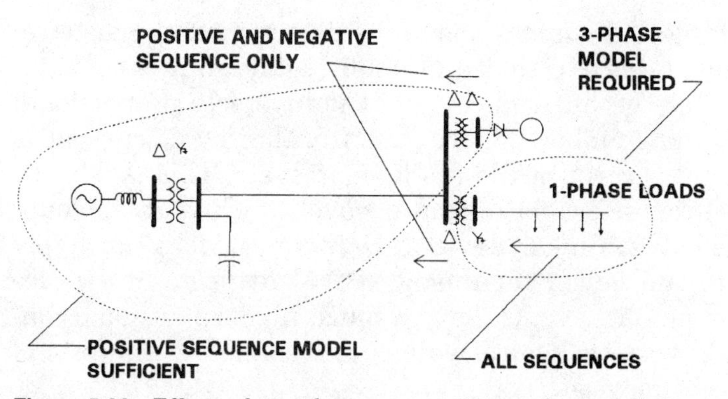

Figure 5.29 Effects of transformer connection on the modeling requirements for analyzing harmonic flows in networks.

programs capable of accurately modeling these systems simply set up the coupled three-phase equations and solve them. It takes no more time than to solve the sequence networks because they would have to be coupled also. Not only does the technique fail to yield an advantage, but analysts also begin to make errors and inadvertently violate the assumptions of the method. It is recommended that the technique be avoided by those who are not absolutely certain of their understanding when performing unbalanced circuit analysis.

In summary, many harmonic cases can be analyzed using familiar symmetrical component modeling techniques. In the case of three-phase industrial loads, nearly all such loads can be analyzed using the positive-sequence impedance model. The most notable exceptions are harmonics from single-phase loads on utility distribution feeders and 120/208 V circuits in industrial and commercial buildings.

5.18 Modeling Harmonic Sources

Most harmonic analysis is performed using steady-state, linear circuit solution techniques. Harmonic sources, which are nonlinear elements, are generally considered to be injection sources into the linear network models.

For most harmonic flow studies, it is suitable to treat harmonic sources as simple sources of harmonic currents. This is the nominal case for power system devices when the voltage distortion at the service bus is generally relatively low, less than 5 percent. This is illustrated in Fig. 5.30, where an electronic power converter is replaced with a current source in the equivalent circuit.

Values of injected current should be determined by measurement. In the absence of that and published data, it is common to assume that the harmonic content is inversely proportional to the harmonic number. That is, the fifth-harmonic current is one-fifth or 20 percent of the fundamental, etc. This is derived from the Fourier series for a square wave, which is at the foundation of many nonlinear devices. However, it does not apply very well to the newer-technology PWM drives and switch-mode power supplies, which have a much higher harmonic content. Table 5.4 shows typical values to assume for analysis of several types of devices.

When the system is near resonance, a simple current source model will give an excessively high prediction of voltage distor-

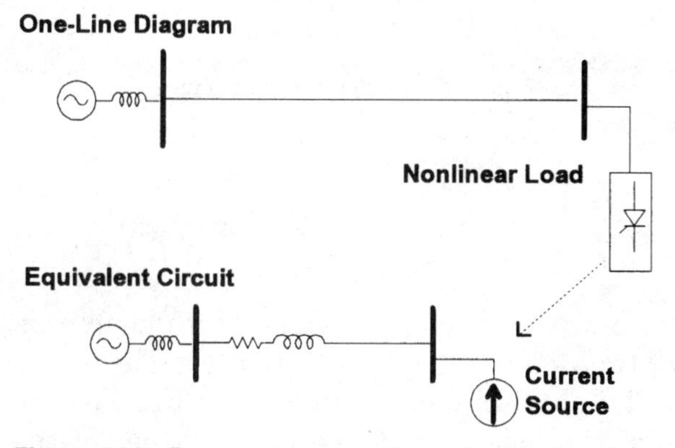

Figure 5.30 Representing a nonlinear load with a harmonic current source for analysis.

TABLE 5.4 Typical Percent Harmonic Distortion of Common Harmonic Sources: Odd Harmonics, 1 through 13

Harmonic	6-pulse ASD	PWM drive	Arc lighting	SMPS
1	100	100	100	100
3			20*	70
5	18	90	7	40
7	12	80	3	15
9			2.4*	7
11	6	75	1.8	5
13	4	70	0.8	3

ASD, adjustable-speed drive; PWM, pulse-width modulated; SMPS, switch-mode power supply.
*For single-phase or unbalanced three-phase modeling; otherwise, assume triplen is zero.

tion. The model tries to inject a constant current into a high impedance, which is not a valid representation of reality. Often, this is inconsequential because the most important thing is to know that the system cannot be successfully operated in resonance, which is readily observable from the simple model. Once the resonance is eliminated by, e.g., adding a filter, the model will give a realistic answer.

For the cases where a more accurate answer is required during resonant conditions, a more sophisticated model must be used. For many power system devices, a Thevenin or Norton equivalent is adequate (Fig. 5.31). The additional impedance modifies the response of the parallel resonant circuit.

A Thevenin equivalent is obtained in a straightforward manner for many nonlinear loads. For example, an arc furnace is well

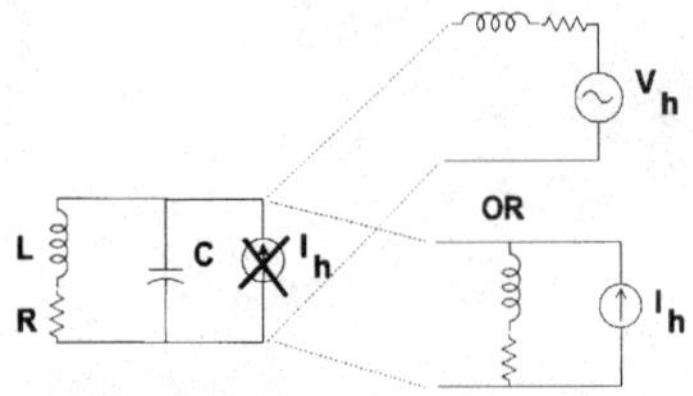

Figure 5.31 Replacing simple current source model with a Thevinen or Norton equivalent for better source models of resonant conditions.

represented by a square-wave voltage of peak magnitude approximately 50 percent of the nominal ac system voltage. The series impedance is simply the short-circuit impedance of the furnace transformer and leads (the lead impedance is the larger of the two). Unfortunately, it is difficult to determine clear-cut equivalent impedances for many nonlinear devices. In these cases, a detailed simulation of the internals of the harmonic-producing load is necessary. This can be done with computer programs that iterate on the solution or through detailed time-domain analysis.

Fortunately, it is seldom essential to obtain such great accuracy during resonant conditions and analysts do not often have to take these measures. However, we usually model arcing devices with a Thevenin model.

5.19 Harmonic Filter Design

Harmonic filter design will be illustrated through a simple, but common, example. A single-tuned 480-V notch filter, illustrated in Fig. 5.32, will be designed for the fifth harmonic. The filter is tuned slightly below the harmonic frequency of concern. This method allows for tolerances in the filter components and prevents the filter from acting as a direct short circuit for the offending harmonic current. This allows the filter to perform its function while helping to reduce the duty on the filter components. It also minimizes the possibility of dangerous harmonic resonance should the system parameters change and cause the tuning frequency to shift slightly higher.

The general method for applying filters is as follows:

- Apply one single-tuned shunt filter first and design it for the lowest generated frequency.
- Determine the voltage distortion level at the low-voltage bus.
- Vary the filter elements according to the specified tolerances and check the filter's effectiveness.

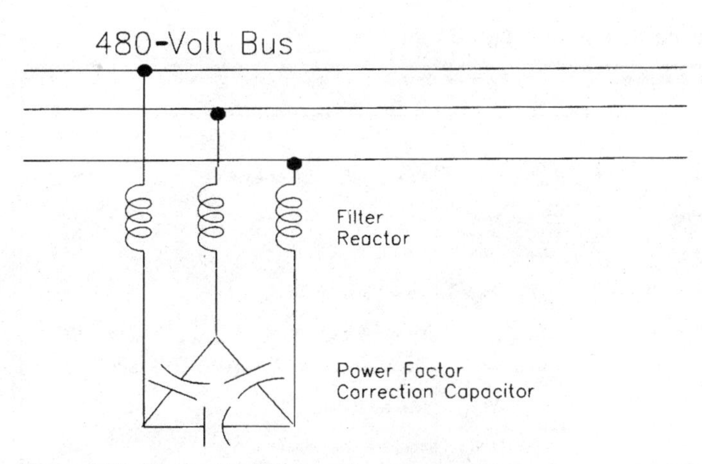

Figure 5.32 Example of a low-voltage configuration.

- Check the frequency response characteristic to verify that the newly created parallel resonance is not close to a harmonic frequency.
- If necessary, investigate the need for several filters, such as fifth and seventh, or third, fifth, and seventh.

Table 5.5 shows the results of a filter design procedure using a computer spreadsheet. The methods used in this spreadsheet are described in the following:

The actual fundamental frequency compensation provided by a derated capacitor bank is determined using

$$\text{kvar}_{\text{actual}} = \text{kvar}_{\text{rated}} \left(\frac{\text{kV}_{\text{actual}}}{\text{kV}_{\text{rated}}} \right)^2 \tag{5.30}$$

In this case, the rated and actual voltages are the same, so the actual kvar of the capacitor is the rated kvar, 500 kvar. The fundamental frequency current for the capacitor bank is

$$I_{\text{FL}_{\text{cap}}} = \frac{\text{kvar}_{\text{actual}}}{\sqrt{3}\text{kV}_{\text{actual}}} = \frac{500}{\sqrt{3} \times 0.480} = 601.4 \text{ A} \tag{5.31}$$

The equivalent single-phase impedance of the capacitor bank is

TABLE 5.5 Harmonic Filter Design Example

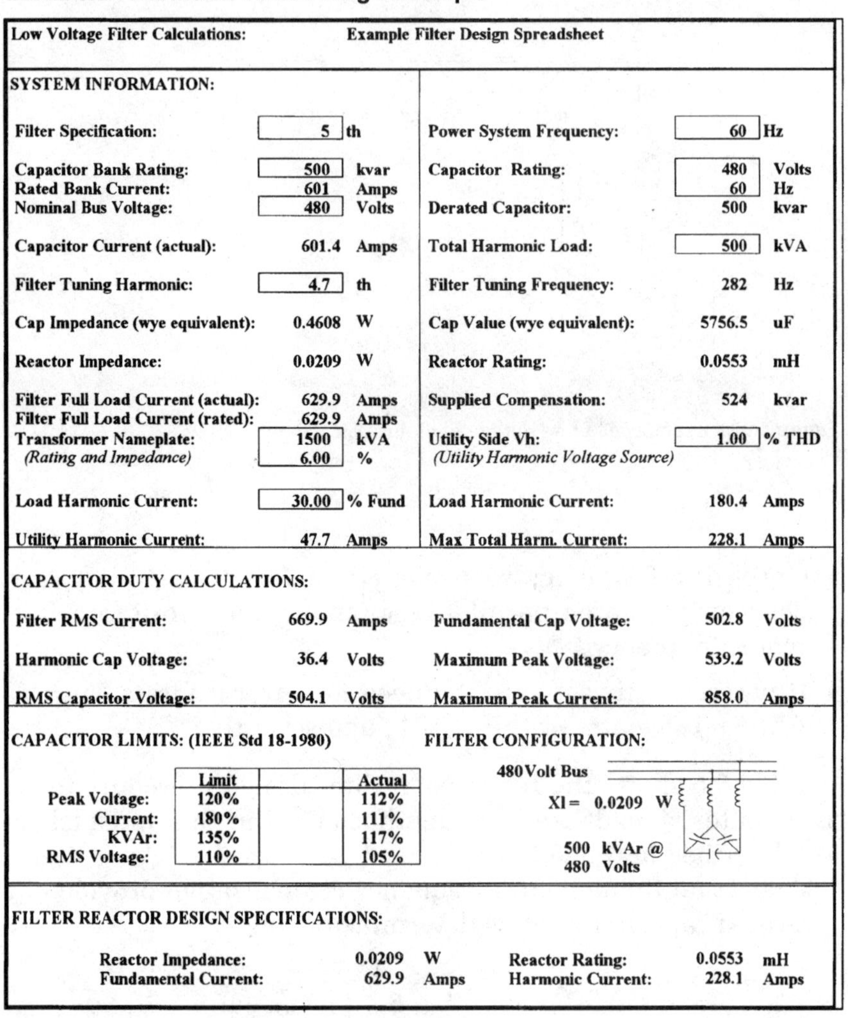

Low Voltage Filter Calculations: **Example Filter Design Spreadsheet**

SYSTEM INFORMATION:

Filter Specification:	5	th	Power System Frequency:	60	Hz
Capacitor Bank Rating:	500	kvar	Capacitor Rating:	480	Volts
Rated Bank Current:	601	Amps		60	Hz
Nominal Bus Voltage:	480	Volts	Derated Capacitor:	500	kvar
Capacitor Current (actual):	601.4	Amps	Total Harmonic Load:	500	kVA
Filter Tuning Harmonic:	4.7	th	Filter Tuning Frequency:	282	Hz
Cap Impedance (wye equivalent):	0.4608	W	Cap Value (wye equivalent):	5756.5	uF
Reactor Impedance:	0.0209	W	Reactor Rating:	0.0553	mH
Filter Full Load Current (actual):	629.9	Amps	Supplied Compensation:	524	kvar
Filter Full Load Current (rated):	629.9	Amps			
Transformer Nameplate:	1500	kVA	Utility Side Vh:	1.00	% THD
(Rating and Impedance)	6.00	%	*(Utility Harmonic Voltage Source)*		
Load Harmonic Current:	30.00	% Fund	Load Harmonic Current:	180.4	Amps
Utility Harmonic Current:	47.7	Amps	Max Total Harm. Current:	228.1	Amps

CAPACITOR DUTY CALCULATIONS:

Filter RMS Current:	669.9	Amps	Fundamental Cap Voltage:	502.8	Volts
Harmonic Cap Voltage:	36.4	Volts	Maximum Peak Voltage:	539.2	Volts
RMS Capacitor Voltage:	504.1	Volts	Maximum Peak Current:	858.0	Amps

CAPACITOR LIMITS: (IEEE Std 18-1980)

	Limit		Actual
Peak Voltage:	120%		112%
Current:	180%		111%
KVAr:	135%		117%
RMS Voltage:	110%		105%

FILTER CONFIGURATION:

FILTER REACTOR DESIGN SPECIFICATIONS:

Reactor Impedance:	0.0209	W	Reactor Rating:	0.0553	mH
Fundamental Current:	629.9	Amps	Harmonic Current:	228.1	Amps

$$X_{C_Y} = \frac{\text{kV}^2_{\text{rated}}}{\text{Mvar}_{\text{rated}}} = \frac{0.480^2}{0.5} = 0.4608\ \Omega \tag{5.32}$$

The filter reactor impedance is determined using

$$X_R = \frac{X_C}{n^2} = \frac{0.4608\ \Omega}{4.7^2} = 0.02086\ \Omega \tag{5.33}$$

Including the filter reactor increases the fundamental current to

$$I_{\mathrm{FL}_{\mathrm{filter}}} = \frac{V_{\mathrm{bus}}}{\sqrt{3}(X_{\mathrm{C}}+X_{\mathrm{R}})} = \frac{0.480}{\sqrt{3}(-0.4608 + 0.0209)} = 629.9 \text{ A} \tag{5.34}$$

Due to the fact that the filter draws more fundamental current than the capacitor alone, the supplied kvar compensation is larger than the capacitor rating and can be determined using

$$\mathrm{kvar}_{\mathrm{supplied}} = \sqrt{3} \times V_{\mathrm{bus}} \times I_{\mathrm{FL}_{\mathrm{filter}}} = \sqrt{3} \times 480 \times 629.9 = 524 \text{ kvar} \tag{5.35}$$

Capacitor ratings should be compared with standard capacitor limits as shown at the bottom of Table 5.5. Filter reactor specifications should include both a fundamental and harmonic current value. The harmonic current should be determined assuming a reasonable value for background distortion from other sources. In this case, it was assumed that the utility side voltage distortion was 1.0 percent.

The tuning characteristic of the filter is described by its quality factor, Q. Q is a measure of the sharpness of tuning and, for series filter resistance, is defined as

$$Q = \frac{nX_{\mathrm{L}}}{R} \tag{5.36}$$

where R = series resistance of filter elements
n = tuning harmonic
X_{L} = reactance of filter reactor at fundamental frequency

Typically, the value of R consists of only the resistance of the inductor. This usually results in a very large value of Q and a very sharp filtering action. This is normally satisfactory for the typical single-filter application and results in a filter that is very economical to operate (small energy consumption). However, sometimes it is desirable to introduce some intentional losses to help dampen the response of the system. A resistor is commonly added in *parallel* with the reactor to create a high-pass filter. In this case, Q is defined as the inverse of Eq. (5.36) so that large numbers reflect sharp tuning. High-pass filters are generally used only at the 11th and 13th harmonics, and higher. It is usually not economical to operate such a filter at the fifth and seventh harmonics because of the amount of losses and the size of the resistor.

The reactors used for larger filter applications are generally built with an air core, which provides linear characteristics with respect to frequency and current. Reactors for smaller filters and filters that must fit into a confined space or near steel structures are built with a steel core. A ± 5 percent tolerance in the reactance is usually acceptable for industrial applications. The 60-Hz X/R ratio is usually between 50 and 150. A series resistor may be used to lower this ratio, if desired, to produce a filter with more damping. The reactor should be rated to withstand a short circuit between the reactor and capacitor. A design Q for the high-pass configuration might typically be 1 or 2 to achieve a flat response above the tuned frequency.

Filters for many high-power, three-phase applications such as static var systems almost always include fifth and seventh harmonics because those are the largest harmonics produced by the six-pulse bridge. Occasionally this causes a system resonance near the third harmonic that may require a third-harmonic filter. Normally, one wouldn't think that the third harmonic would be a problem in a three-phase bridge, but imbalances in the operation of the bridge and in system parameters create small amounts of uncharacteristic harmonics. If the system responds to those harmonics, filters will have to be applied anyway.

5.20 Telecommunications Interference

Harmonic currents flowing on the utility distribution system or within an end-user facility can create interference in communication circuits sharing a common path. Voltages induced in parallel conductors by the common harmonic currents often fall within the bandwidth of normal voice communications. Harmonics between 540 Hz (ninth harmonic) and 1200 Hz are particularly disruptive. The induced voltage per ampere of current increases with frequency. Triplen harmonics (3rd, 9th, 15th) are especially troublesome in four-wire systems because they are in phase in all conductors of a three-phase circuit and, therefore, add directly in the neutral circuit, which has the greatest exposure with the communications circuit.

Harmonic currents on the power system are coupled into communication circuits by either induction or direct conduction. Figure 5.33 illustrates coupling from the neutral of an

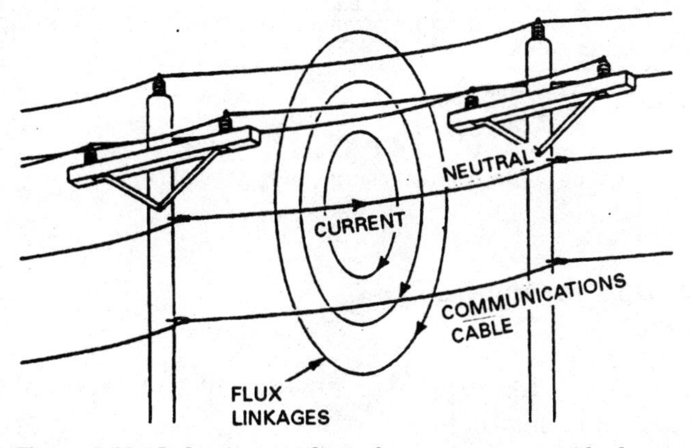

Figure 5.33 Inductive coupling of power system residual current to telephone circuit.

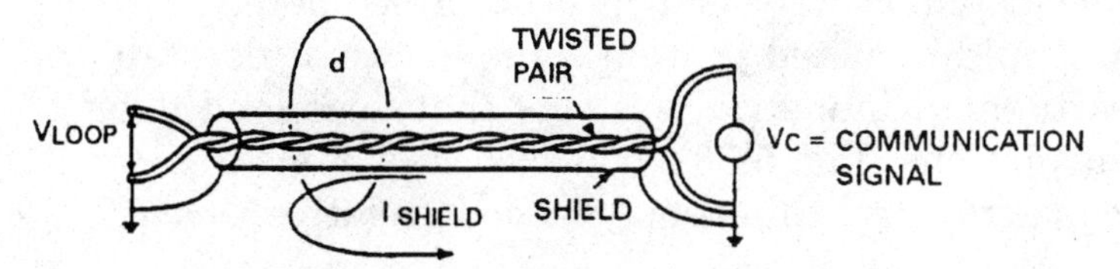

Figure 5.34 IR drop in cable shield resulting in potential differences in ground references at ends of cable.

overhead distribution line by induction. This was a severe problem in the days of open wire telephone circuits. Now, with the prevalent use of shielded, twisted-pair conductors for telephone circuits, this mode of coupling is less significant. The direct inductive coupling is equal in both conductors, resulting in zero net voltage in the loop formed by the conductors.

Inductive coupling can still be a problem if high currents are induced in the shield surrounding the telephone conductors. Current flowing in the shield causes an IR drop (Fig. 5.34), which results in a potential difference in the ground references at the ends of the telephone cable.

Shield currents can also be caused by direct conduction. As illustrated in Fig. 5.35, the shield is in parallel with the power system ground path. If local ground conditions are such that a relatively large amount of current flows in the shield, high-shield IR drop will again cause a potential difference in the ground references at the ends of the telephone cable.

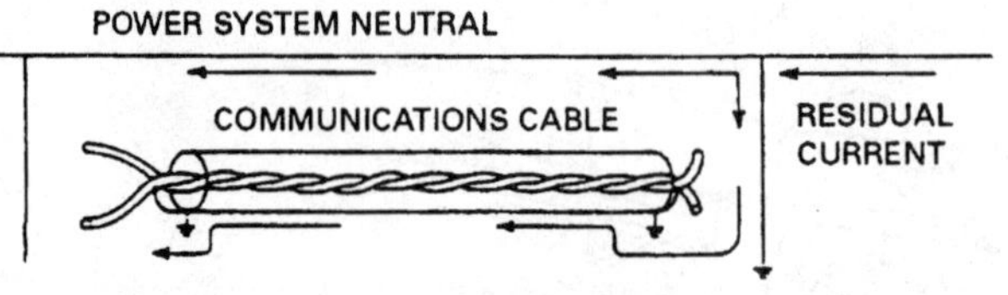

Figure 5.35 Conductive coupling through a common ground path.

5.21 Computer Tools for Harmonics Analysis

The preceding discussion has provided an idea of the types of functions that must be performed for harmonics analysis of power systems. It should be rather obvious that for anything but the simplest of circuits, a sophisticated computer program is required. The characteristics of such programs and the heritage of some popular analysis tools is described here.

First, it should be noted that one simple circuit does appear frequently in small industrial systems that does lend itself to manual calculations (Fig. 5.36). It is basically a one-bus circuit with one capacitor. Two things may be done relatively easily:

1. Determine the resonant frequency. If the resonant frequency is near a potentially damaging harmonic, either the capacitor must be changed or a filter designed.

2. Determine an estimate of the voltage distortion due to the current, I_h. The voltages, V_h, are given by

$$V_h = \left(\frac{R + j\omega L}{1 - \omega^2 LC + j\omega RC} \right) I_h \qquad h = 2, 3, \ldots \tag{5.37}$$

$$\omega = 2\pi f_1 h$$

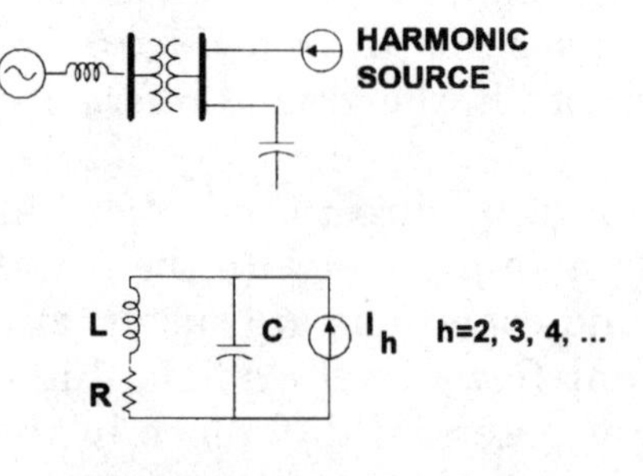

Figure 5.36 A simple circuit which may be analyzed manually.

Given that the resonant frequency is not near a significant harmonic and that projected voltage distortion is low, the application will probably be successful.

Unfortunately, not all practical cases can be represented with such a simple circuit. In fact, adding just one more bus with a capacitor to the simple circuit in Fig. 5.36 makes the problem a real challenge to even the most skilled analysts. However, a computer can perform the chore in milliseconds.

To use the computer tools commonly available, the analyst must describe the circuit configuration, loads, and the sources to the program. Data that must be collected include

- Line and transformer impedances
- Transformer connections
- Capacitor values and locations (critical)
- Harmonic spectra from nonlinear loads
- Power source voltages

These values are entered into the program, which automatically adjusts impedances for frequency and computes the harmonic flow throughout the system.

5.21.1 Capabilities for harmonics analysis programs

Acceptable computer software for harmonics analysis of power systems should have the following characteristics:

1. It should be capable of handling large networks of at least several hundred nodes.
2. It should be capable of handling multiphase models of arbitrary structure. Not all circuits, particularly those on utility distribution feeders are amenable to accurate solution by balanced, positive-sequence models.
3. It should be capable of modeling systems with positive-sequence models. When there can be no zero-sequence harmonics, there is no sense in wasting computer resources for a full three-phase model.
4. It should be able to perform frequency scan at small intervals of frequency (e.g., 10 Hz) to develop the system frequency response characteristics necessary to identify resonances.

5. It should be able to perform simultaneous solution of numerous harmonic sources to estimate the actual current and voltage distortion.
6. It should have built-in models of common harmonic sources.
7. It should allow both current source and voltage source models of harmonic sources.
8. It should be able to automatically adjust phase angles of the sources based on the fundamental frequency phase angles.
9. It should be able to model any transformer connection.
10. It should be able to display the results in a meaningful and friendly manner to the user.

5.21.2 Harmonic analysis by computer—historical perspective

While general-purpose circuit analysis programs can be used, there have been several computer programs developed specifically for analysis of harmonic flows in power systems. These programs perform harmonic flow analysis in the steady state. The analysis could also be performed in the time domain, but this is generally more time-consuming and is generally excessive for most problems.

It is informative to review the history of harmonics analysis that has led to some of the major software packages that are available today. Since this area of analysis is less well known than the field of load flow analysis, it also gives us an opportunity to acknowledge the contributions of several of the pioneers in this field, especially those whose contributions are not obvious from the published literature.

Prior to the widespread use of computers for harmonics analysis, power systems harmonic studies were frequently performed on analog simulators such as a transient network analyzer (TNA). The few TNAs in the United States in the mid-1970s were located at large equipment manufacturers, primarily at General Electric Co., Westinghouse Electric, and McGraw-Edison Power Systems. Because of the inconvenience and high cost, harmonic studies were generally performed only on very special cases such as large arc furnace installations that might impact utility transmission systems. TNAs usually had at least two variable-frequency sources. Therefore, the general procedure was to use one source for the power frequency and the second source to represent the nonlinear load, one frequency at a time. One tricky part of this procedure was to sweep the frequency through system resonances fast enough to avoid burning up the power supply or damaging inductors and capacitors.

Our involvement in harmonics analysis began in 1975, when author Roger Dugan, then with McGraw-Edison, constructed the first electronic arc model for a TNA to eliminate the need to use the second source.[3] In that same year, to overcome the limitation of harmonic analysis by analog simulator, Dugan teamed up with Dr. Sarosh N. Talukdar and William L. Sponsler at McGraw-Edison to develop one of the first commercial computer programs specifically designed to automate analysis of harmonic flows on large-scale power systems. Dubbed the Network Frequency Response Analysis Program (NFRAP), it was developed for the Virginia Electric and Power Company to study the impacts of adding 220-kV capacitor banks on the transmission system. Prior to that, engineers typically used a transients program or modified versions of short-circuit analysis programs, often modifying the impedances by hand for the different frequencies.

The NFRAP program techniques, which are direct nodal admittance matrix solution techniques that treat nonlinearities as sources, evolved into what is probably the most prolific family of harmonic analysis programs. From 1977 to 1979, EPRI sponsored an investigation of harmonics on utility distribution feeders.[5,6] One of the products of this research was the Distribution Feeder Harmonics Analysis program. It was the first program designed specifically to analyze harmonics on unbalanced distribution systems and had specific models of power systems elements to help the user develop models. It became the prototype for the modern harmonic analysis program. Key investigators on this project, RP 1024-1, were Robert E. Owen and author Mark McGranaghan, and the key software designers were again Dugan and Sponsler.

The next generation of software tool based on the NFRAP program methodology was the McGraw-Edison Harmonic Analysis Program (MEHAP), under development from 1980 to 1984. It was written in FORTRAN for minicomputers and had the distinction of being interactive with graphical output. All previous efforts had been batch mode programs with tabular output.

For its time, it fit the definition of being "user friendly." The developers included Dugan, McGranaghan, and Jack A. King.

At about this same time, however, the PC revolution took place. Erich W. Gunther recoded the algorithms in the Pascal language and created the V-HARM® program.[4] To the authors' knowledge, this was the first commercial harmonic analysis program written expressly for the PC environment. It has proven to be a very reliable and durable program and can be licensed today from Cooper Power Systems. Gunther subsequently has written the latest generation in this heritage of harmonic analysis tools in the C++ language for the Microsoft® Windows™ environment. It is called the SuperHarm™ program and can be licensed from Electrotek Concepts, Inc.

The CYMHARMO® program was developed first at Institut de Recherche d'Hydro-Quebec (IREQ) in Montreal in 1983 and now can be licensed through CYME International, Inc. The program was originally written in FORTRAN for the mainframe and was ported to the PC shortly afterward in 1984. It is now written in a mixture of FORTRAN and C languages. The principal authors of the software are Dr. Chinh Nguyen and Dr. Ali Moshref. The Canadian Electric Association (CEA) has supported the development of this program, which uses analysis techniques similar to the previously mentioned programs.

Since 1981, EPRI has sponsored the development of the HARMFLO program, which takes a different approach to the network solution. Drs. G. L. Heydt, D. Xia, and W. Mack Grady[1,7] developed the program at Purdue University and based it on the Newton-Raphson power flow techniques. The program was the first to adjust the harmonic current output of the load for the harmonic voltage distortion. The FORTRAN program was originally developed for mainframe batch computers and is now also available on the PC. The latest version is 5.0 and can be licensed by eligible parties from EPRI.

Of course, harmonics problems can be solved on transients analysis programs such as the EMTP, originally developed by the Bonneville Power Administration. The special-purpose programs are generally more efficient for normal problems, but occasionally, a very difficult problem will be encountered that requires simulation in the time domain.

Nowadays, harmonic analysis of one sort or another is a common feature of many software vendors' packages for both utility and industrial power system analysis. What used to be a very exotic type of analysis may very well become commonplace to the next generation of power engineers.

5.22 References

1. D. Xia and G. T. Heydt, "Harmonic Power Flow Studies: Part I—Formulation and Solution," *IEEE Transactions on Power Apparatus and Systems,* June 1982, pp. 1257–1265.
2. J. M. Frank, "Origin, Development and Design of K-Factor Transformers," in *Conference Record,* 1994 IEEE Industry Applications Society Annual Meeting, Denver, October 1994, pp. 2273–2274.
3. R. C. Dugan, "Simulation of Arc Furnace Power Systems," *IEEE Transactions on Industry Applications,* November/December 1980, pp. 813–818.
4. M. F. McGranaghan and E. W. Gunther, "Design of a PC-Based Harmonic Simulation Program," in *Proceedings of the Second International Conference on Harmonics in Power Systems,* Winnipeg, Manitoba, October 1986.
5. M. F. McGranaghan, J. H. Shaw, and R. E. Owen, "Measuring Voltage and Current Harmonics on Distribution Systems, *IEEE Transactions on Power Apparatus and Systems,* Vol. 101, No. 7, July 1981.
6. M. F. McGranaghan, R. C. Dugan, and W. L. Sponsler, "Digital Simulation of Distribution System Frequency Response Characteristics," *IEEE Transactions on Power Apparatus and Systems,* Vol. 101, No. 3, March 1981.
7. W. M. Grady, "Harmonic Power Flow Studies," Ph.D. thesis, Purdue University, May 1983.

5.23 Bibliography

The material presented in this chapter is based on the following publications:

Dugan, R. C., McGranaghan, M. F., Rizy, D. T., and Stovall, J. P, *Electric Power System Harmonics Design Guide,* ORNL/Sub/81-95011/3, Oak Ridge National Laboratory, U.S. Department of Energy, September 1987.

Dwyer, R. V., Gunther, E. W., and Adapa, R., "A Comparison of Solution Techniques for the Calculation of Harmonic Distortion due to Adjustable Speed DC Drives," in *Fourth International Conference on Harmonic Systems,* Budapest, Hungary, October 1990.

Grebe, T. E., McGranaghan, M. F., and Samotyj, M., "Solving Harmonic Problems in Industrial Plants and Harmonic Mitigation Techniques for Adjustable-Speed Drives, in *Proceedings of Electrotech 92,* Montreal, 1992.

IEEE Standard C57.110-1986, *IEEE Recommended Practice for Establishing Transformer Capability When Supplying Nonsinusoidal Load Currents* (reaffirmed 1992), Piscataway, N.J., 1986.

IEEE Standard 18-1992, *IEEE Standard for Shunt Power Capacitors*, Piscataway, N.J., 1992.

IEEE Standard 519-1992, *IEEE Recommended Practice and Requirements for Harmonic Control in Electric Power Systems*, Piscataway, N.J., 1992.

McGranaghan, M. F., and Mueller, D. R., "Designing Harmonic Filters for Adjustable-Speed Drives to Comply with New IEEE-519 Harmonic Limits," in *Proceedings of the IEEE/IAS Annual Conference (Petroleum and Chemical Industry Technical Conference)*, 1993.

McGranaghan, M. F., Grebe, T. E., and Samotyj, M. "Solving Harmonic Problems in Industrial Plants—Case Studies," in *Proceedings of the First International Conference on Power Quality (PQA '91)*, Paris, 1991.

Schwabe, R. J., Melhorn, C. J., and Samotyj, M., "Effect of High Efficiency Lighting on Power Quality in Public Buildings," in *Proceedings of the Third International Conference on Power Quality (PQA '93)*, San Diego.

Zavadil, R., McGranaghan, M. F., Hensley, G., and Johnson, K., "Analysis of Harmonic Distortion Levels in Commercial Buildings," in *Proceedings of the First International Conference on Power Quality (PQA '91)*, Paris 1991.

Chapter

6

Long-Duration Voltage Variations

Utilities generally try to maintain the service voltage supplied to an end user within ±5 percent of nominal. Under emergency conditions, for short periods, ANSI Standard C84.1 permits the utilization voltage to be in the range +6 percent to −13 percent of the nominal voltage. Some sensitive loads have more stringent voltage limits for proper operation and, of course, equipment generally operates more efficiently at near-nominal voltage. This chapter addresses the fundamental problems behind voltage regulation and the general types of devices available to correct the problem.

6.1 Principles of Regulating the Voltage

The root cause of most voltage regulation problems is that there is too much impedance in the power system to properly supply the load (Fig. 6.1). Therefore, the voltage drops too low under heavy load. Conversely, when the source voltage is boosted to overcome the impedance, there can be an overvoltage condition when the load drops too low. The corrective measures usually involve either compensating for the impedance, Z, or compensating for the voltage drop, $IR + jIX$, caused by the impedance.

The options for improving the voltage regulation are

1. Add voltage regulators, which boosts the apparent V_1.
2. Add shunt capacitors to reduce the current, I, and shift it to be more in phase with the voltage.

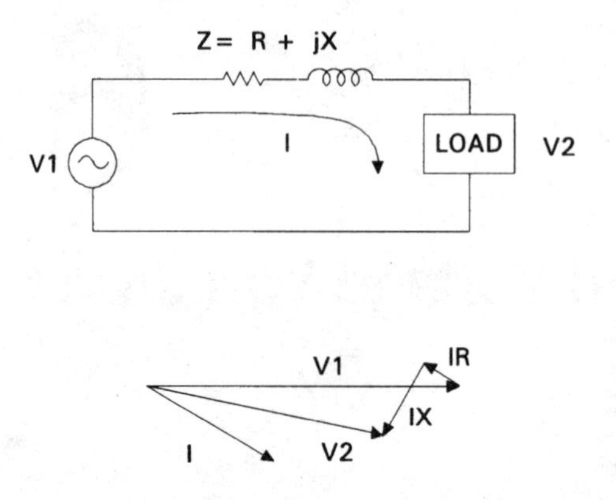

Figure 6.1 Voltage drop across the system impedance is the root cause of voltage regulation problems.

3. Add series capacitors to cancel the inductive impedance drop (*IX*).
4. Reconductor lines to a larger size to reduce the impedance, *Z*.
5. Change the service transformer to a larger size to reduce impedance, *Z*.
6. Add static var compensators, which serve the same purpose as capacitors for rapidly changing loads.

6.2 Devices for Voltage Regulation

There are a variety of voltage regulation devices in use on utility and industrial power systems. We typically divide these devices into three major classes:

1. Tap-changing transformers
2. Isolation devices with separate voltage regulators
3. Impedance compensation devices, such as capacitors

There are both mechanical and electronic tap-changing transformers. Most of the tap-changing transformer designs are autotransformers, although there are also numerous applications of two- and three-winding transformers with tap changers. The mechanical devices are for the slower-changing loads while the electronic ones can respond very quickly to voltage changes.

Isolation devices include UPS systems, ferroresonant (constant voltage) transformers, M-G sets, and the like. These are devices that essentially isolate the load from the power source by performing some sort of energy conversion. Therefore, the load side of the device can be separately regulated and can maintain constant voltage regardless of what is occurring at the power supply. The downside of using such devices is that they introduce more losses and may also cause harmonics problems on the power supply system.

Shunt capacitors help maintain the voltage by reducing the current in the lines. Also, by overcompensating inductive circuits, a voltage rise can be achieved. To maintain a more constant voltage, the capacitors can be switched in conjunction with the load, sometimes in small incremental steps to follow the load more closely. If the objective is simply to maintain the voltage at a higher value to avoid an undervoltage condition, the capacitors are often fixed (not switched).

Series capacitors are relatively rare, but are useful for some impulse loads like rock crushers and tire testers.[1] Many potential users shy away from them because of the extra care in engineering required for the series capacitor installation to function properly. However, they are very effective in certain system conditions, primarily with rapidly changing large loads that are causing excessive flicker.

The series capacitors compensate for most of the inductance in the system leading up to the load. If the system is highly inductive, this represents a significant reduction in the impedance. If the system is not highly inductive, but has a high proportion of resistance, series capacitors will not be very effective. This is typical of many industrial systems that have long lengths of cable between the transformer and the load. Reconductoring or changing the transformer must be done to achieve a significant reduction in the impedance.

Another approach to flicker-causing loads is to apply static var compensators. These can react within a few cycles to maintain a fairly constant voltage by controlling the reactive power production. Such devices are commonly used on arc furnaces and other randomly varying loads where the system is weak and the resulting flicker is affecting nearby customers.

6.2.1 Utility step-voltage regulators

The typical utility tap-changing regulator can regulate from −10 to +10 percent of the incoming line voltage in 32 steps of ⅝ percent. There are some variations, but the majority are of this type. Distribution substation transformers commonly have three-phase load tap changers (LTCs) while regulators in-

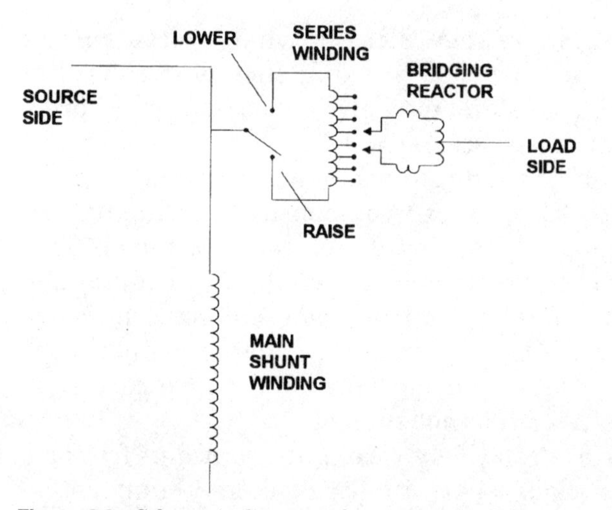

Figure 6.2 Schematic diagram of one type of utility voltage regulator commonly applied on distribution lines.

stalled out on the feeders are typically single-phase. Line regulators may be installed in banks of two or three; it is not uncommon to have open-delta banks on three-phase feeders with light to moderate load for purposes of economy.

Figure 6.2 shows a schematic of a utility step voltage regulator. Although the concept of a tap-changing autotransformer is simple, a utility voltage regulator is a fairly complicated piece of apparatus designed to achieve a long life and high reliability of the tap-changing mechanism.

Utility voltage regulators are relatively slow. The time delay for when the voltage goes out of band is at least 15 s and is commonly 30 or 45 s. Thus, it is not suitable where voltages may vary in matters of cycles or seconds. Their main application is boosting voltage on long feeders. The voltage band typically ranges from 1.5 to 3.0 V on a 120-V base. The control can be set to maintain voltage at some point downline from the feeder by using the *line drop compensator.* This results in a more level average voltage response and helps prevent overvoltages for customers near the regulator.

6.2.2 Ferroresonant transformers

On the end-user side, ferroresonant transformers are not only useful in protecting equipment from voltage sags, but they can also be used to attain very good voltage regulation (±1 percent output). Figure 6.3 shows the steady-state input/output charac-

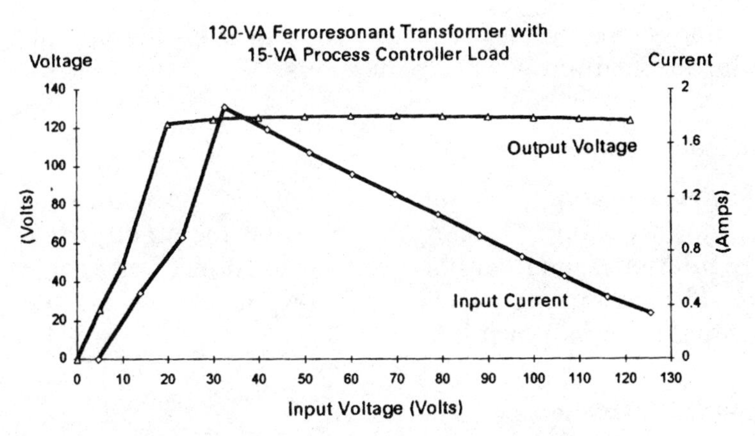

Figure 6.3 Ferroresonant transformer steady-state characteristics.

teristics of a 120-VA ferroresonant transformer with a 15-VA load. As the input voltage is reduced down to 30 V, the output voltage stays constant. If the input voltage is reduced further, the output voltage begins to collapse. In addition, as the input voltage is reduced, the current drawn by the ferroresonant transformer increases substantially from 0.4 to 2 A. However, ferroresonant transformers tend to be lossy and inefficient.

6.2.3 Electronic tap-switching regulator

An electronic tap-switching regulator (Fig. 6.4) can also be used to regulate voltage. They are more efficient than ferroresonant transformers, and use silicon-controlled rectifiers (SCRs) or triacs to quickly change taps, and hence voltage. Tap-switching

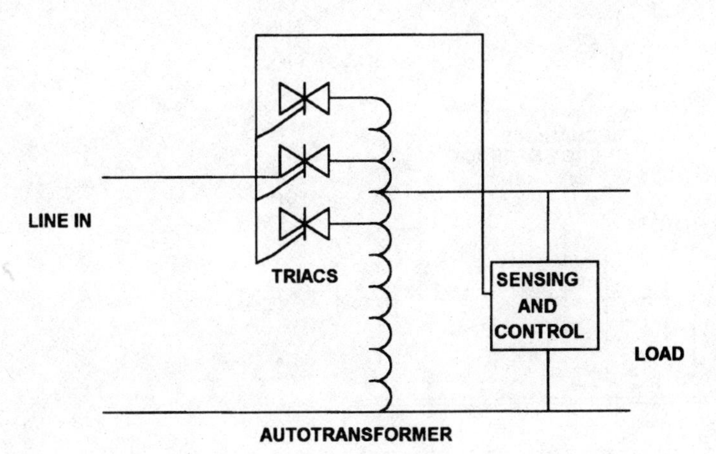

Figure 6.4 Electronic tap-switching regulator.

regulators have a very fast response time of a half-cycle, and are popular for medium-power applications.

6.2.4 Magnetic synthesizers

Magnetic synthesizers, although intended for short-duration voltage sags (see Chap. 3), can also be used for steady-state voltage regulation. One manufacturer, for example, states that for input voltages of ±40 percent, the output voltage will remain within ±5 percent at full load.

6.2.5 On-line UPS systems

On-line UPS systems intended for protection against sags and brief interruptions can also be used for voltage regulation provided the source voltage stays sufficiently high to keep the batteries charged. This is a common solution for small, critical computer or electronic control loads in an industrial environment that has large, fluctuating loads causing the voltage to vary.

6.2.6 Motor-generator sets

Motor-generator sets (Fig. 6.5) are also used for voltage regulation. They completely decouple the load from the electric power system, shielding the load from transients. Voltage regulation is provided by the generator control. The major drawback of M-G sets is their response time to large load changes. Motor-generator sets can take several seconds to bring the voltage back up to the required level, making this device too slow for voltage regu-

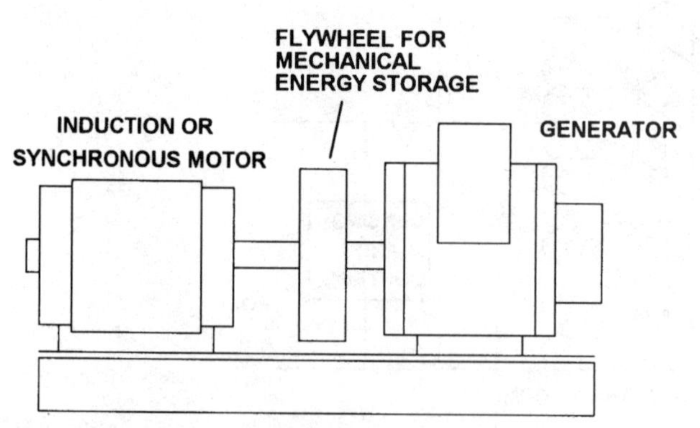

Figure 6.5 Motor-generator set.

lation of certain loads, especially rapidly varying loads. Motor-generator sets can also be used to provide ride through from input voltage variations, especially voltage sags, by storing energy in a flywheel.

6.2.7 Static var compensators

Static var compensators can be applied either to utility systems or to industrial systems. They help regulate the voltage by responding very quickly to supply or consume reactive power. This acts with the system impedance to either raise or lower the voltage on a cycle-by-cycle basis.

There are two main types of static var compensator in common usage, as shown in Fig. 6.6. The thyristor-controlled reactor (TCR) scheme is probably the most common. It employs a fixed-capacitor bank to provide leading reactive power and a thyristor-controlled inductance that is gated on in various amounts to cancel all or part of the effect of the capacitance. The capacitors are frequently configured as filters to clean up the harmonic distortion caused by the thyristors.

The thyristor-switched capacitor operates by switching multiple steps of capacitors quickly to match the load requirements as closely as possible. This is a more coarse regulation than a TCR, but is often adequate. The capacitors are generally gated fully on so there are no harmonics in the currents. The switching point is controlled so that there are no switching transients.

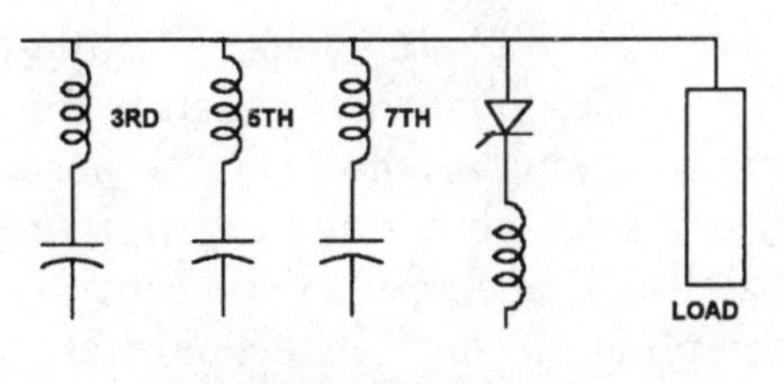

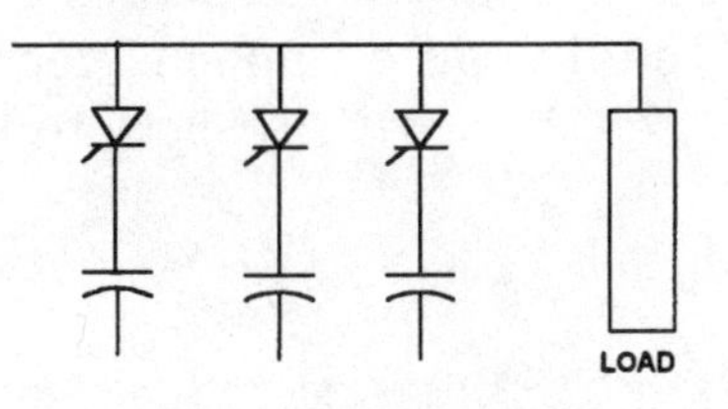

Figure 6.6 Common static var compensator configuration.

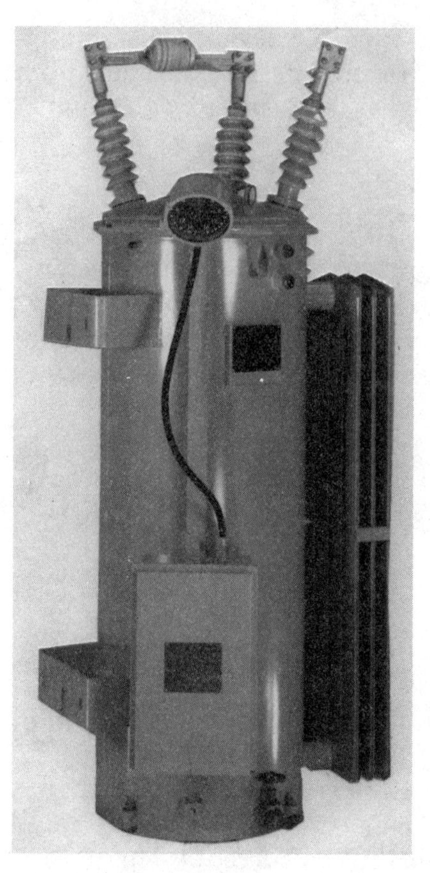

Figure 6.7 Typical utility 32-step voltage regulator. (*Courtesy of Cooper Power Systems.*)

6.3 Utility Voltage Regulator Application

Figure 6.7 shows a photograph of a typical 32-step voltage regulator used by U.S. utilities. This is a single-phase device that is frequently pole-mounted, either one to a single pole or three on a platform between two poles. They may be connected in wye-grounded, leading delta, lagging delta, or open delta. The controls are integral to the device and each phase is controlled separately.

Volumes could be written on the application of regulators, but we will restrict our discussion here to a few topics particularly relevant to power quality: use of the line drop compensator for leveling voltage profiles and load rejection with respect to the application of regulators in series.

6.3.1 Line drop compensator

Regulators are very effective in alleviating low-voltage conditions on distribution feeders when the load has outgrown the capability of the feeder at peak load conditions. Because it is time-consuming to determine the correct settings for line drop compensation, the *R* and *X* settings are often set to zero and the voltage regulation set point is set near the maximum allowable (125 or 126 V on a 120-V base). This results in the feeder voltage being near the maximum most of time because the load is at peak for only a small percentage of the hours each year. This is satisfactory in most respects except that

1. Transformers operate higher on their saturation curve, producing more harmonic currents (and losses), contributing more to the harmonic distortion on the feeder, which can be particularly troublesome at low loads.
2. Customers may experience more frequent replacement of incandescent lamps.

The purpose of the line drop compensator is to level out the voltage profile so that it provides the necessary voltage boost at peak load yet keeps the voltage closer to nominal at lower loads. This is illustrated in Fig. 6.8. To simplify the discussion, we've assumed there is no LTC in the substation and the only regulator of concern is a feeder regulator at the substation. In part *a,* no compensation is used and the voltage setting is 5 percent high, or 126 V on 120-V systems. Since there is some bandwidth on the control, the voltage may actually go higher than this. In part *b,* the voltage setting is 120 V (100 percent) with the line drop compensator set some distance out on the feeder as shown. At peak load the voltage at the regulator rises to 105 percent, which is necessary to keep the end of the feeder at the proper voltage. However, at low load, the feeder voltage profile is closer to 100 percent voltage.

There are numerous practices for determining line drop compensator settings. Manufacturers provide computer programs for computing the settings provided the CT and PT ratios are known. These vary with regulator sizes and must be specifically known before the proper setting can be computed. Of course, this also requires the user to model the feeders on a computer program, for which the data may not be readily available. Manufacturer's guide books also have simple formulas and rules-of-thumb for determining settings.

The line drop compensator settings are called *R* and *X* for the resistive and reactive portions of the compensator, respectively. However, the units are volts on a 120-V base instead of ohms.

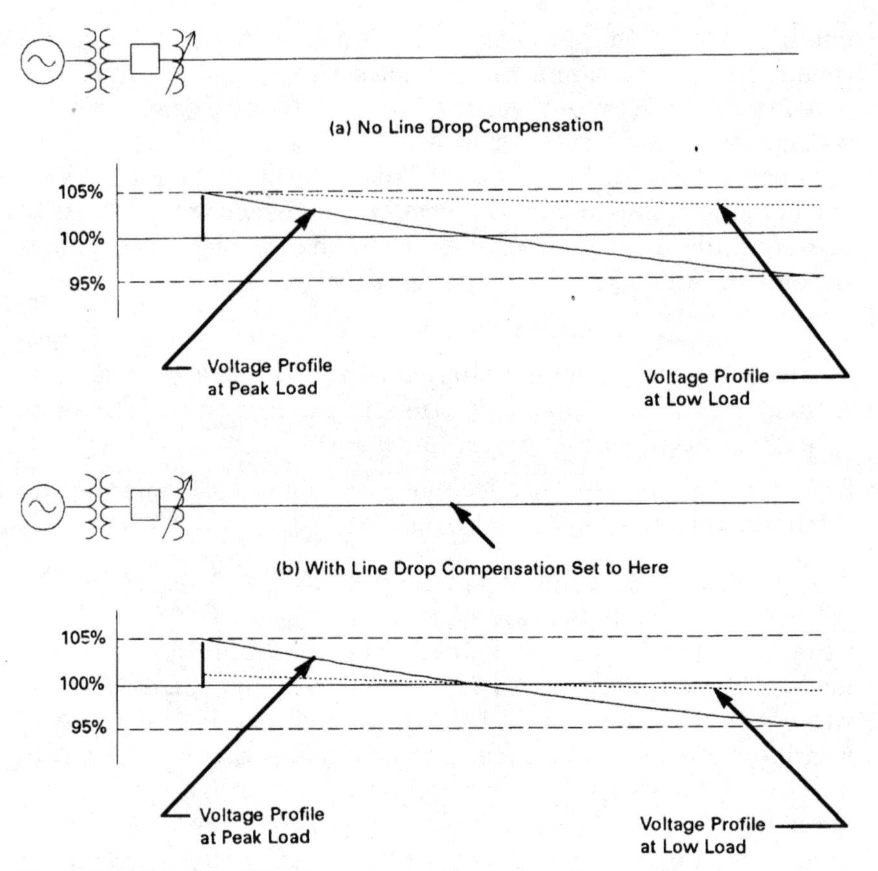

Figure 6.8 The effect of line drop compensation on the voltage profile.

To convert from actual line impedance in ohms to the *R* and *X* settings, the basic formula is

$$(R + jX)_{\text{setting}} = (R + jX)_{\text{ohms}}\left(\frac{\text{CT rating}}{\text{PT ratio}}\right) \tag{6.1}$$

where the CT is specified by the line current *rating* and the PT *ratio* is the nominal line-neutral voltage divided by 120 V.

These *R* and *X* values are used directly for wye-connected regulators. For delta-connected regulators, these values must be modified to account for the 30-degree phase shift in the voltage with respect to the line current. For a leading delta connection, multiply by $1\angle{-30}$ degrees, for a lagging delta, multiply by $1\angle{+30}$ degrees.

Some utilities have developed average standard settings that they have found to be effective. Many determine the *R* and *X* settings experimentally by sending a line technician to the low-voltage point on the feeder while another adjusts the R and X settings. Ideally, this should be done at the peak load so that a voltage setting and line drop setting may be found that is successful in meeting this condition. It will, in all likelihood, meet the lower load conditions satisfactorily, although downline switched capacitor banks may fool the control somewhat when they switch to a different state. Therefore, the voltage profile should be monitored at one or two key locations for a few days to make certain the setting is adequate.

Obviously, this process takes time and it is often not convenient to send a crew to check a regulator setting when the peak load occurs. Often, at this time, the crews are busy with more urgent matters such as changing out overloaded transformers to get customers back in service. There is a definite benefit to the power quality if the regulator is set properly, so some effort should be made. Fortunately, manufacturers are now supplying controls with telecommunications capability so that the settings can be adjusted more conveniently from a control center.

Many manufacturers also offer sophisticated controls with a choice of load following algorithms. In the case of power quality complaints with the voltage going out of band or too many tap changes, consult the user's manual and experiment with other algorithms to achieve a smoother regulation.

6.3.2 Regulators in series

In sparsely populated areas it is not uncommon to find two or more regulator banks in series on extremely long lines feeding remote loads. Two notable applications are service to irrigation and mining loads where lines extend for miles with only an occasional load. These applications require special considerations to avoid power quality problems.

One important consideration for coordinating the regulators in series is properly setting the initial time delay. The regulator nearest the substation is set with the shortest time delay, typically 15 or 30 s. Regulators further downline are set with time delay of 15 s longer. This minimizes tap changing on the downline regulators, keeping the voltage variations to a minimum, and extending contact life.

Perhaps the greatest power quality problem in this situation is load rejection. The sudden loss of load, which can happen

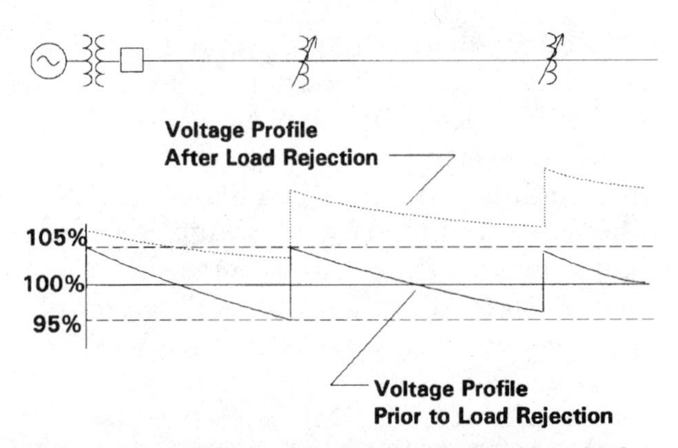

Figure 6.9 Illustration of overvoltage resulting from load rejection on regulators in series.

after a fault, will result in greatly excessive voltages because the regulator boosting will be cumulative (Fig. 6.9). Overvoltages of 20 percent or more can occur. Transformer saturation and remaining load will help hold the voltage down, but it will still exceed normal limits by a considerable margin.

To minimize damage to loads, regulators employ a "rapid runback" control scheme that bypasses the normal time delay and runs the regulators back down as quickly as possible. This is typically 2 to 4 s per tap change.

6.4 Capacitors for Voltage Regulation

Capacitors may be used for voltage regulation on the power system in either the shunt or series configuration.

6.4.1 Shunt capacitors

As shown in Fig. 6.10*a*, the presence of a shunt capacitor at the end of a feeder results in a gradual change in voltage along the feeder. Ideally, the percent voltage rise at the capacitor

$$\%\Delta V = \frac{100 \cdot (V_{\text{with cap}} - V_{\text{no cap}})}{V_{\text{with cap}}} \tag{6.2}$$

would be zero at no load, and rise to maximum at full load. However with shunt capacitors, percent voltage rise is essentially independent of load. Therefore, automatic switching is often employed in order to deliver the desired regulation at high loads, but prevent excessive voltage at low loads. This

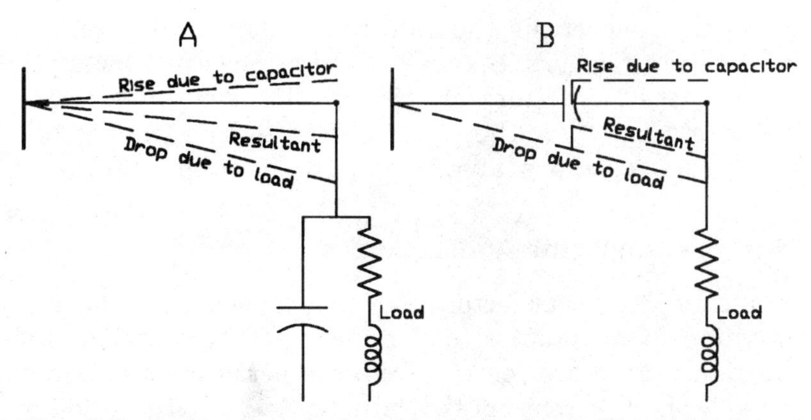

Figure 6.10 Feeder voltage rise due to shunt (*a*) and series (*b*) capacitors.

practice may result in transient overvoltages inside customer facilities, as described in Chap. 4.

Application of shunt capacitors may also result in a variety of harmonic problems (see Chap. 5).

6.4.2 Series capacitors

Unlike the shunt capacitor, a capacitor connected in series with the feeder results in a voltage rise at the end of the feeder that varies directly with load current. Voltage rise is zero at no load, and maximum at full load. Thus, series capacitors do not need to be switched in response to changes in load. Moreover, a series capacitor requires far smaller kilovolt and kvar ratings than a shunt capacitor delivering equivalent regulation.

However, series capacitors have several disadvantages. First, they cannot provide reactive compensation for feeder loads, and so do not significantly reduce system losses. Series capacitors can only release additional system capacity if it is limited by excessive feeder voltage drop. Shunt capacitors, on the other hand, are also effective when system capacity is limited by high feeder current.

Second, series capacitors cannot tolerate fault current. This would result in a catastrophic overvoltage, and must be prevented by bypassing the capacitor through an automatic switch. An arrester must also be connected across the capacitor to divert current until the switch closes.

There are several other concerns that must be evaluated in a series capacitor application. These include resonance and/or hunting with synchronous and induction motors, and ferroresonance with transformers. Because of these concerns, the application of series capacitors on distribution systems is very limited. One area where they have proved to be advantageous is where feeder reactance must be minimized to reduce flicker.

6.5 End-User Capacitor Application

The application of power factor correction capacitors is generally motivated by economics to eliminate utility power factor penalties, but there are reasons from the perspective of power quality as well. The reasons that an end user might decide to apply power factor correction capacitors are

- To reduce the electric utility bill
- To reduce I^2R losses and, therefore, heating in lines and transformers
- To increase the voltage at the load, increasing production and/or the efficiency of the operation
- To reduce current in the lines and transformers, allowing additional load to be served without building new circuits

There can be power quality problems as the result of adding capacitors. The most common are harmonics problems. While power factor correction capacitors are not harmonic sources, they can interact with the system to accentuate the harmonics that are already there (see Chap. 5). There are also switching transient side effects such as magnification of utility capacitor switching transients (see Chap. 4).

6.5.1 Location for power factor correction capacitors

The benefits realized by installing power factor correction capacitors include the reduction of reactive power flow on the system. Therefore, for best results, power factor correction should be located as close to the load as possible. However, this may not be the most economical solution or even the best engineering solution, due to the interaction of harmonics and capacitors.

Often capacitors are installed with large induction motors (C3 in Fig. 6.11). This allows the capacitor and motor to be switched as a unit. Large plants with extensive distribution

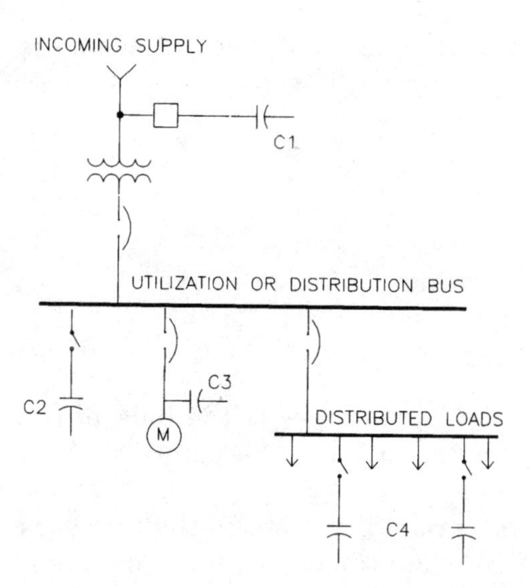

Figure 6.11 Location of power factor correction.

systems often install capacitors at the primary voltage bus (C1) when utility billing encourages power factor correction. Many times however, power factor correction and harmonic distortion reduction must be accomplished with the same capacitors. Location of larger harmonic filters on the distribution bus (C2) provides the required compensation and a low-impedance path for harmonic currents to flow, keeping the harmonic currents off the utility system.

The disadvantage of placing capacitors only at the utilization or distribution bus is that there is no reduction of current and line losses within the plant. Loss and current reduction is achieved when the capacitors are distributed throughout the system. Some industrial end users install capacitors at the motor control centers, which is often more economical than putting the capacitors on each motor. The capacitors' controls can be tied in with the motor controls so that the capacitors are switched when needed.

6.5.2 Voltage rise

The voltage rise from placing capacitors on an inductive circuit is a two-edged sword from the power quality standpoint. If the voltage is low, then the capacitors provide an increase to bring the voltage back into tolerable limits. However, if the capacitors are left energized when the load is turned off, the voltage can rise too high, resulting in a sustained overvoltage.

The voltage rise realized with the installation of capacitors is approximated from

$$\% \Delta V = \frac{\text{kvar}_{\text{cap}} \times Z_{\text{tx}} (\%)}{\text{kVA}_{\text{tx}}} \tag{6.3}$$

where $\% \Delta V$ = percent voltage rise
kvar_{cap} = capacitor bank rating
kVA_{tx} = step-down transformer rating
Z_{tx} = step-down transformer impedance, %

This formula assumes that the transformer is the bulk of the total impedance of the power system up to the point at which the capacitor is applied.

As mentioned above, one power quality problem that arises is that the voltage rises too high when the capacitors remain energized at low load levels. One common symptom of this is loud humming in the supply transformer and, in some cases, overheating due to overexcitation of the core. Another symptom is the loss of excessive numbers of incandescent light bulbs coincident with the installation of a capacitor bank. Thus, this formula should be applied to investigate whether it is feasible to leave the capacitors energized. If not, some control strategy must be devised to switch the capacitors off at light loads.

6.5.3 Reduction in power system losses

Since losses are inversely proportional to the power factor squared (PF^2), the reduction in power system losses is estimated from

$$\% \text{ power loss} \propto 100 \left(\frac{\text{PF}_{\text{original}}}{\text{PF}_{\text{corrected}}} \right)^2 \tag{6.4}$$

$$\% \text{ loss reduction} = 100 \left[1 - \left(\frac{\text{PF}_{\text{original}}}{\text{PF}_{\text{corrected}}} \right)^2 \right] \tag{6.5}$$

where % loss reduction = percent reduction in losses
$\text{PF}_{\text{original}}$ = original power factor (pu)
$\text{PF}_{\text{corrected}}$ = corrected power factor (pu)

This formula basically applies to a single capacitor on a radial feed. However, it is also approximately correct if the capacitors are well distributed throughout the plant so that each major branch circuit experiences approximately the same percentage loss improvement.

Keep in mind that this formula gives the percent reduction possible over the present losses *upline* from the capacitors. There is no reduction in losses in the lines and transformers between the capacitor and the load.

6.5.4 Reduction in line current

The percent line current reduction can be approximated from

$$\%\Delta I = 100\left[1 - \left(\frac{\cos\,\theta_{\text{before}}}{\cos\,\theta_{\text{after}}}\right)\right] \tag{6.6}$$

where $\%\Delta I$ = percent current reduction
θ_{before} = power factor angle before correction
θ_{after} = power factor angle after correction

Again, this applies only to currents upline from the capacitor.

6.5.5 Displacement power factor vs. true power factor

The traditional concepts of selecting power factor correction are based on the assumption that loads on the system have linear voltage-current characteristics and that harmonic distortion can be ignored. With these assumptions, the power factor is equal to the *displacement power factor* (DPF). DPF is calculated using the traditional power factor triangle method (Fig. 6.12) and is often written:

$$\text{DPF} = \frac{\text{kW}}{\text{kVA}} = \cos\theta \tag{6.7}$$

where kW and kVA are the fundamental frequency quantities only.

Harmonic distortion in the voltage and current caused by nonlinear loads on the system changes the way power factor

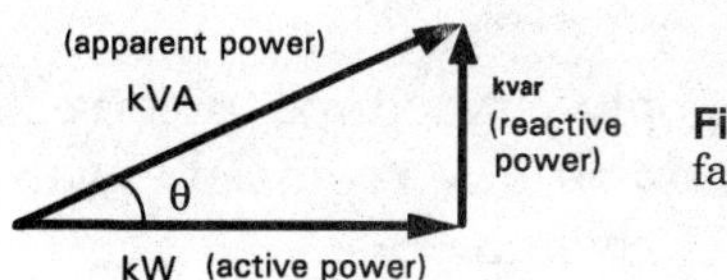

Figure 6.12 Displacement power factor triangle.

must be calculated. *True power factor* (TPF) is defined as the ratio of real power to the total volt-amperes in the circuit:

$$\text{TPF} = \frac{\text{kW}}{\text{kVA}} = \frac{P}{V_{\text{rms}} \times I_{\text{rms}}} \qquad (6\text{-}8)$$

As before, the power factor is defined as the ratio of kW to kVA, but in this case, the kVA includes harmonic distortion volt-amperes. The total kVA (apparent power) is determined by multiplying the true rms voltage by the true rms current. It can be significantly higher than the fundamental frequency kVA. The active power, P, is generally increased only marginally by the distortion.

TPF is the true measure of the efficiency with which the real power is being used. In the trivial case of no distortion, it defaults to the DPF. Capacitors basically compensate only for the fundamental frequency reactive power (vars) and cannot completely correct the true power factor to unity when there are harmonics present. In fact, capacitors can make true power factor worse by creating resonance conditions which magnify the harmonic distortion. On typical power systems, the I_{rms} term in the above equation is generally the one most affected by harmonic distortion although the V_{rms} term may also be increased.

Assuming the voltage THD is zero, the maximum to which you can correct the true power factor can be approximated by

$$\text{TPF} \approx \sqrt{\frac{1}{1 + \text{THD}^2_{\text{current}}}} \qquad (6.9)$$

(THD in pu). DPF is still very important to most industrial customers because utility billing for power factor penalties is generally based on it. Most revenue metering schemes currently account only for the DPF. However, this could change because modern electronic meters certainly have the capability to compute the TPF, which will be considerably lower for some types of industrial loads.

6.5.6 Selecting the amount of capacitance

For reference for those wishing to apply capacitors to correct the power factor, the kvar rating of capacitance required to correct a load to a desired power factor is given by

$$\text{kvar} = \text{kW}(\tan\phi_{\text{orig}} - \tan\phi_{\text{new}})$$

$$= \sqrt{\frac{1}{\text{PF}^2_{\text{orig}}} - 1} - \sqrt{\frac{1}{\text{PF}^2_{\text{new}}} - 1} \qquad (6.10)$$

where kvar = required compensation in kvar
kW = real power in kW
ϕ_{orig} = original power factor phase angle
ϕ_{new} = desired power factor phase angle
PF_{orig} = original power factor
PF_{new} = desired power factor

Table 6.1 summarizes the above equation in tabular form.

After selecting estimated capacitor sizes, two power quality checks should be done:

1. Determine the no-load voltage rise to make sure that the voltage will not rise above 110 percent when the load is minimum. If it does, you will have to switch some of the capacitors off or apply fewer capacitors.
2. Determine the impact of the capacitors on harmonics (see Chap. 5).

If harmonics prove to be a problem, typical options are:

1. Change the amount of capacitors, if possible. Avoid certain switching combinations. This is generally the least expensive solution.
2. Convert some of the capacitors to one or more filters, usually placed at the main bus.
3. Employ an adaptive control to monitor the harmonic distortion and switch the capacitors to avoid resonance. This might be appropriate for large industrial loads where numerous switched capacitors are coming on and off line randomly.

6.6 Regulating Utility Voltage with Dispersed Sources

It is becoming more popular for distribution planners to consider dispersed generation and storage devices to postpone investments in substations and transmission lines until the load has grown to a sufficient size to warrant the investment. This con-

TABLE 6.1 kW Multiplier to Determine kvar Requirement

Original PF	0.80	0.82	0.84	0.86	0.88	0.90	0.92	0.94	0.96	0.98	1.00
0.50	0.982	1.034	1.086	1.139	1.192	1.248	1.306	1.369	1.440	1.529	1.732
0.52	0.893	0.945	0.997	1.049	1.103	1.158	1.217	1.280	1.351	1.440	1.643
0.54	0.809	0.861	0.913	0.965	1.019	1.074	1.133	1.196	1.267	1.356	1.559
0.56	0.729	0.781	0.834	0.886	0.940	0.995	1.053	1.116	1.188	1.276	1.479
0.58	0.655	0.707	0.759	0.811	0.865	0.902	0.979	1.042	1.113	1.201	1.405
0.60	0.583	0.635	0.687	0.740	0.794	0.849	0.907	0.970	1.042	1.130	1.333
0.62	0.515	0.567	0.620	0.672	0.726	0.781	0.839	0.903	0.974	1.062	1.265
0.64	0.451	0.503	0.555	0.607	0.661	0.716	0.775	0.838	0.909	0.998	1.201
0.66	0.388	0.440	0.492	0.545	0.599	0.654	0.712	0.775	0.847	0.935	1.138
0.68	0.328	0.380	0.432	0.485	0.539	0.594	0.652	0.715	0.787	0.875	1.078
0.70	0.270	0.322	0.374	0.427	0.480	0.536	0.594	0.657	0.729	0.817	1.020
0.72	0.214	0.266	0.318	0.370	0.424	0.480	0.538	0.601	0.672	0.761	0.964
0.74	0.159	0.211	0.263	0.316	0.369	0.425	0.483	0.546	0.617	0.706	0.909
0.76	0.105	0.157	0.209	0.262	0.315	0.371	0.429	0.492	0.563	0.652	0.855
0.78	0.052	0.104	0.156	0.209	0.263	0.318	0.376	0.439	0.511	0.599	0.802
0.80	0.000	0.052	0.104	0.157	0.210	0.266	0.324	0.387	0.458	0.547	0.750
0.82		0.000	0.052	0.105	0.158	0.214	0.272	0.335	0.406	0.495	0.698
0.84			0.000	0.053	0.106	0.162	0.220	0.283	0.354	0.443	0.646
0.86				0.000	0.054	0.109	0.167	0.230	0.302	0.390	0.593
0.88					0.000	0.055	0.114	0.177	0.248	0.337	0.540
0.90						0.000	0.058	0.121	0.193	0.281	0.484
0.92							0.000	0.063	0.134	0.223	0.426
0.94								0.000	0.071	0.160	0.363
0.96									0.000	0.089	0.292
0.98										0.000	0.203
1.00											0.000

cept is particularly useful when there are a relatively few number of hours each year when the load approaches the system capacity limits. The devices are installed in sizes ranging from 500 kW to 10 MW, and many of them are transportable so that they can be reused at a future date in other locations.

For the present, most of the installations have been considered for the utility distribution substations. This offers load relief for the substation and transmission facilities, but contributes little else to the quality of power for the distribution feeder. Many distribution engineers are now considering the consequences of moving the devices out onto the feeder to gain additional loss reduction, improved reliability, and voltage regulation. While this option may be too expensive to consider for voltage regulation alone, it is a useful consequence of dispersed sources being justified on the basis of deferment of capital expansion.

One potential use of dispersed sources for voltage regulation is related to reliability considerations (Fig. 6.13). Utilities usually have line switches installed so that portions of the distribution feeder can be served from different feeders or substations during emergencies. If the fault occurs at the time of peak load, it may be impossible to pick up any more load from other feeders in the normal manner. However, a generator located close to the switch tie point can potentially provide enough power to support the additional load at a satisfactory voltage.

One advantage of using a generator to regulate the voltage is that its controls generally respond faster and more smoothly than discrete tap-changing devices like regulators and LTCs.

The controls of dispersed sources must be carefully coordinated with existing line regulators and substation LTCs. With conventional regulators, reverse power flow can sometimes fool the

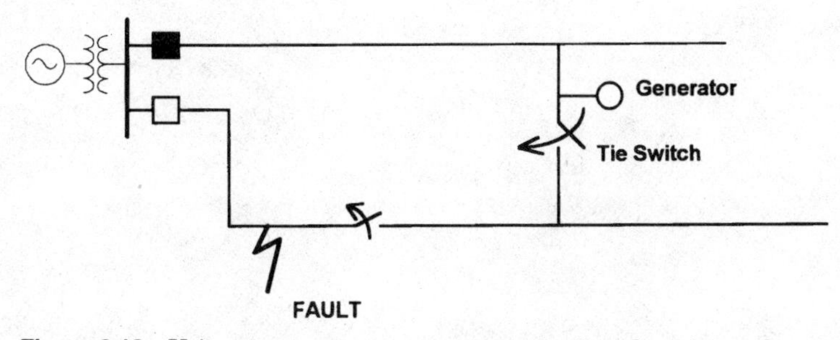

Figure 6.13 Using a generator to support restoration of service to the unfaulted portion of a feeder.

regulators into moving the tap changer in the wrong direction. Also, it is possible for the generator to cause regulators to change taps constantly, causing early failure of the tap-changing mechanism. Fortunately, some regulator manufacturers have anticipated these problems and now provide sophisticated microcomputer-based regulator controls that are able to compensate.

To exploit dispersed sources for voltage regulation, the options are limited to the types of sources with steady, controllable outputs such as gas engines and battery storage. Randomly varying sources such as wind turbines and photovoltaic cells are unsatisfactory for this role and often must be placed on a relatively stiff part of the system or have special regulation to compensate for their output variations.

6.7 Reference

1. L. Morgan and S. Ihara, "Distribution Feeder Modification to Service Both Sensitive Loads and Large Drives," in *1991 IEEE PES Transmission and Distribution Conference Record,* Dallas, September 1991, pp. 686–690.

Chapter

7

Wiring and Grounding

Many power quality variations that occur within customer facilities are related to wiring and grounding problems. It is commonly reported at power quality conferences that 80 percent of all the power quality problems reported by customers are related to wiring and grounding problems within a facility. While there may be no scientific data to support this precise percentage, many power quality problems are solved by simply tightening a loose connection or replacing a corroded conductor. Therefore, an evaluation of wiring and grounding practices is a necessary first step when evaluating power quality problems in general.

The *National Electrical Code*® (*NEC*®)* and other important standards provide the minimum standards for wiring and grounding. It is often necessary to go beyond the requirements of these standards to achieve a system which also minimizes the impact of power quality variations (harmonics, transients, noise) on connected equipment. This section provides general information on proper wiring and grounding practices and also outlines common problems that are encountered.

7.1 Definitions

Selected definitions are presented from the *IEEE Dictionary* (Standard 100), the *Green Book* (IEEE Standard 142), and the *National Electrical Code*. Both the *Green Book* and the *NEC* pro-

**National Electrical Code*® and *NEC*® are registered trademarks of the National Fire Protection Association, Inc., Quincy, Mass. 02269.

vide extensive information on proper grounding practices for safety considerations and proper system operation. However, these documents do not address all concerns for power quality. Power quality considerations associated with wiring and grounding practices are covered in Federal Information Processing Standard (FIPS) 94—*Guideline on Electrical Power for ADP Installations.* The *Emerald Book,* currently under development (IEEE Project 1100), will include an update to the information presented in FIPS 94. Grounding guidelines to minimize noise in electronics circuits is covered in IEEE Standard 518, *IEEE Guide for the Installation of Electrical Equipment to Minimize Electrical Noise Inputs to Controllers from External Sources.* EPRI's *Wiring and Grounding for Power Quality* (Publication CU.2026.3.90) also provides an excellent summary of typical wiring and grounding problems along with recommended solutions.

Some of the key definitions of wiring and grounding terms from these documents are included in the following.

From the *IEEE Dictionary**

grounding A conducting connection, whether intentional or accidental, by which an electric circuit or equipment is connected to the earth, or to some conducting body of relatively large extent that serves in place of the earth. It is used for establishing and maintaining the potential of the earth (or of the conducting body) or approximately that potential, on conductors connected to it; and for conducting ground current to and from the earth (or the conducting body).

Green Book definitions†

ungrounded system A system, circuit, or apparatus without an intentional connection to ground, except through potential indicating or measuring devices or other very high impedance devices.

---

*Reprinted from IEEE Standard 100-1992, *IEEE Standard Dictionary of Electrical and Electronic Terms*, copyright © 1993 by the Institute of Electrical and Electronics Engineers, Inc. The IEEE disclaims any responsibility or liability resulting from the placement and use in this publication. Information is reprinted with the permission of the IEEE.

†Reprinted from IEEE Standard 142-1991, *IEEE Recommended Practice for Grounding of Industrial and Commercial Power Systems*, copyright © 1991 by the Institute of Electrical and Electronics Engineers, Inc. The IEEE disclaims any responsibility or liability resulting from the placement and use in this publication. Information is reprinted with the permission of the IEEE.

grounded system A system of conductors in which at least one conductor or point (usually the middle wire or neutral point of transformer or generator windings) is intentionally grounded, either solidly or through an impedance.

grounded solidly Connected directly through an adequate ground connection in which no impedance has been intentionally inserted.

grounded effectively Grounded through a sufficiently low impedance such that for all system conditions the ratio of zero sequence reactance to positive sequence reactance (X0/X1) is positive and less than 3, and the ratio of zero sequence resistance to positive sequence reactance (R0/X1) is positive and less than 1.

resistance grounded Grounded through impedance, the principal element of which is resistance.

inductance grounded Grounded through impedance, the principal element of which is inductance.

NEC* definitions. Refer to Fig. 7.1.

grounding electrode The grounding electrode shall be as near as practicable to and preferably in the same area as the grounding conductor connection to the system. The grounding electrode shall be: (1) the nearest available effectively grounded structural metal member of the structure; or (2) the nearest available effectively grounded metal water pipe; or (3) other electrodes (Sections 250-81 and 250-83) where electrodes specified in (1) and (2) are not available.

grounded Connected to earth or to some conducting body that serves in place of the earth.

grounded conductor A system or circuit conductor that is intentionally grounded (the neutral is normally referred to as the grounded conductor).

grounding conductor A conductor used to connect equipment or the grounded circuit of a wiring system to a grounding electrode or electrodes.

grounding conductor, equipment The conductor used to connect the noncurrent-carrying metal parts of equipment, raceways, and other enclosures to the system grounded conductor and/or the grounding

*Reprinted with permission from NFPA 70-1993, the *National Electrical Code*®, copyright © 1993, National Fire Protection Association, Quincy, Mass. 02269. This reprinted material is not the complete and official position of the National Fire Protection Association, on the referenced subject which is represented only by the standard in its entirety.

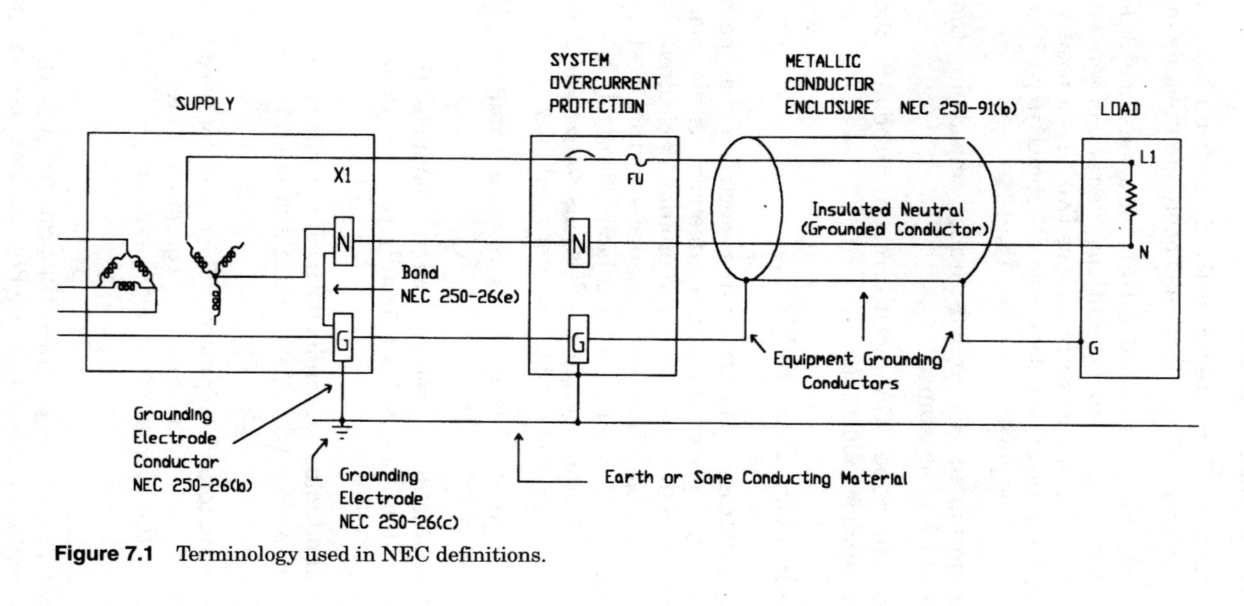

Figure 7.1 Terminology used in NEC definitions.

electrode conductor at the service equipment or at the source of a separately derived system.

grounding electrode conductor The conductor used to connect the grounding electrode to the equipment grounding conductor and/or to the grounded conductor of the circuit at the service equipment or at the source of a separately derived system.

grounding electrode system Defined in *NEC* Section 250-81 as including: (*a*) metal underground water pipe; (*b*) metal frame of the building; (*c*) concrete-encased electrode; and (*d*) ground ring. When these elements are available, they are required to be bonded together to form the grounding electrode system. Where a metal underground water pipe is the only grounding electrode available, it must be supplemented by one of the grounding electrodes specified in Section 250-81 or 250-83.

bonding jumper, main The connector between the grounded circuit conductor (neutral) and the equipment grounding conductor at the service entrance.

branch circuit The circuit conductors between the final overcurrent device protecting the circuit and the outlets.

conduit/enclosure bond (bonding definition) The permanent joining of metallic parts to form an electrically conductive path which will assure electrical continuity and the capacity to conduct safely any current likely to be imposed.

feeder All circuit conductors between the service equipment of the source of a separately derived system and the final branch circuit overcurrent device.

outlet A point on the wiring system at which current is taken to supply utilization equipment.

overcurrent Any current in excess of the rated current of equipment or the capacity of a conductor. It may result from overload, short circuit, or ground fault.

panel board A single panel or group of panel units designed for assembly in the form of a single panel; including buses, automatic overcurrent devices, and with or without switches for the control of light, heat, or power circuits; designed to be placed in a cabinet or cutout box placed in or against a wall or partition and accessible only from the front.

separately derived systems A premises wiring system whose power is derived from generator, transformer, or converter windings and has no direct electrical connection, including a solidly connected grounded circuit conductor, to supply conductors originating in another system.

service equipment The necessary equipment, usually consisting of a circuit breaker switch and fuses, and their accessories, located near the point of entrance of supply conductors to a building or other structure, or an otherwise defined area, and intended to constitute the main control and means of cutoff of the supply.

ufer ground A method of grounding or connection to the earth in which the reinforcing steel (rebar) of the building, especially at the ground floor, serves as a grounding electrode.

7.2 Reasons for Grounding

The most important reason for grounding is safety. There are two important aspects to grounding requirements for safety:

1. *Personnel safety.* Personnel safety is the primary reason that all equipment must have a safety equipment ground. This is designed to prevent the possibility of high touch voltages when there is a fault in a piece of equipment (Fig. 7.2). The touch voltage is the voltage between any two conducting surfaces which can be simultaneously touched by an individual. The earth may be one of these surfaces.

There should be no "floating" panels or enclosures in the vicinity of electric circuits. In the event of insulation failure or inadvertent application of moisture, any electric charge which appears on a panel, enclosure, or raceway must be drained to "ground" or to an object which is reliably grounded.

2. *Grounding to ensure protective device operation.* A ground fault return path to the point where the power source neutral conductor is grounded is an essential safety feature. The *NEC* and some local wiring codes permit electrically continuous conduit and wiring device enclosures to serve as this

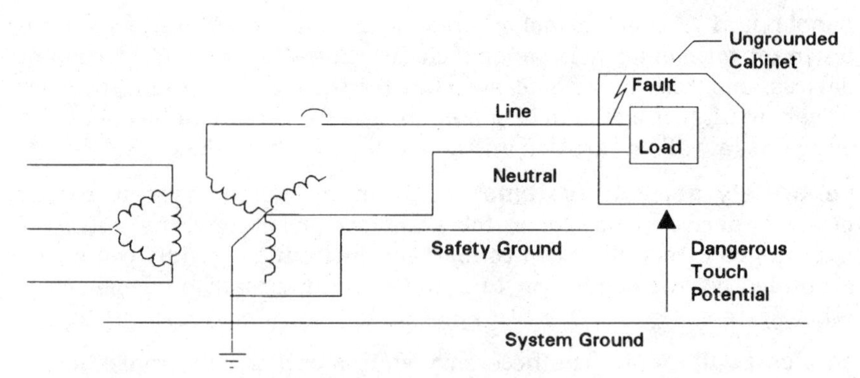

Figure 7.2 High-touch voltage created by improper grounding.

ground return path. Some codes require the conduit to be supplemented with a bare or insulated conductor included with the other power conductors.

An insulation failure or other fault which allows a phase wire to make contact with an enclosure will find a low impedance path back to the power source neutral. The resulting overcurrent will cause the circuit breaker or fuse to disconnect the faulted circuit promptly.

NEC Article 250-51 states that an effective grounding path (the path to ground from circuits, equipment, and conductor enclosures) shall:

a. Be permanent and continuous.

b. Have capacity to conduct safely any fault current likely to be imposed on it.

c. Have sufficiently low impedance to limit the voltage to ground and to facilitate the operation of the circuit protective devices in the circuit.

d. The earth shall not be used as the sole equipment ground conductor.

3. *Noise control.* Noise control includes transients from all sources. This is where grounding relates to power quality. Grounding for safety reasons defines the minimum requirements for a grounding system. Anything that is done to the grounding system to improve the noise performance must be done in addition to the minimum requirements defined in the *NEC* and local codes.

The primary objective of grounding for noise control is to create an equipotential ground system. Potential differences between different ground locations can stress insulation, create circulating ground currents in low-voltage cables, and interfere with sensitive equipment that may be grounded in multiple locations.

Ground voltage equalization of voltage differences between parts of an automated data processing (ADP) grounding system is accomplished in part when the equipment-grounding conductors are connected to the grounding point of a single power source. However, if the equipment-grounding conductors are long, it is difficult to achieve a constant potential throughout the grounding system, particularly for high-frequency noise. Supplemental conductors, ground grids, low-inductance ground plates, etc. may be needed for improving the power quality. These must be in addition to the equipment ground conductors which are required for safety, and not a replacement for them.

7.3 Typical Wiring and Grounding Problems

The following subsections describe some typical power quality problems that are due to inadequacies in the wiring and grounding of electrical systems. It is useful to be aware of these typical problems when performing site surveys because many of the problems can be detected through simple observations. Other problems require measurements of voltages, currents, or impedances in the circuits.

7.3.1 Problems with conductors and connectors

One of the first things to be done during a site survey is to inspect the service entrance, main panel, and major subpanels for problems with conductors or connections. A bad connection (faulty, loose, or resistive connection) will result in heating, possible arcing, and burning of insulation. Table 7.1 summarizes some of the wiring problems that can be uncovered during a site survey.

7.3.2 Missing safety ground

If the safety ground is missing, a fault in the equipment from the phase conductor to the enclosure results in line potential on the exposed surfaces of the equipment. No breakers will trip and a hazardous situation results (Fig. 7.2).

TABLE 7.1 Problems with Conductors and Connectors

Problem observed	Possible cause
Burnt smell at the panel, junction box, or load equipment	Faulted conductor, bad connection, arcing, or overloaded wiring
Panel or junction box is warm to the touch	Faulty circuit breaker or bad connection
Buzzing (corona effect)	Arcing
Scorched insulation	Overloading wiring, faulted conductor, or bad connection
No voltage or load equipment	Tripped breaker, bad connection, or faulted conductor
Intermittent voltage at the load equipment	Bad connection or arcing
Scorched panel or junction box	Bad connection, faulted conductor

7.3.3 Multiple neutral-to-ground connections

Unless there is a separately derived system, the only neutral-to-ground bond should be at the service entrance. The neutral and ground should be kept separate at all panel boards and junction boxes. Downline neutral-to-ground bonds result in parallel paths for the load return current where one of the paths becomes the ground circuit. This can cause misoperation of protective devices. Also, during a fault condition, the fault current will split between the ground and the neutral, which could prevent proper operation of protective devices (a serious safety concern). This is a direct violation of the *NEC*.

7.3.4 Ungrounded equipment

Isolated grounds are sometimes used due to the perceived notion of obtaining a "clean" ground. The proper procedure for using an isolated ground must be followed (see Sec. 7.4.5). Procedures which involve having an illegal insulating bushing in the power source conduit and replacing the prescribed equipment grounding conductor with one to an "Isolated Dedicated Computer Ground" are dangerous, violate code, and are unlikely to solve noise problems.

7.3.5 Additional ground rods

Ground rods for a facility should be part of a grounding system, connected where all the building grounding electrodes (building steel, metal water pipe, etc.) are bonded together. Multiple ground rods can be bused together at the service entrance to reduce the overall ground resistance. Isolated grounds can be used for sensitive equipment, as described above. However, these should not include isolated ground rods to establish a new ground reference for the equipment. One very important power quality problem with additional ground rods is that they create additional paths for lightning stroke currents to flow. With the ground rod at the service entrance, any lightning stroke current reaching the facility goes to ground at the service entrance and the ground potential of the whole facility rises together. With additional ground rods, a portion of the lightning stroke current will flow on the building wiring (green ground conductor and/or conduit) to reach the additional ground rods. This creates a possible transient voltage problem for equipment and a possible overload problem for the conductors.

7.3.6 Ground loops

Ground loops are one of the most important grounding problems in many commercial and industrial environments that include data processing and communication equipment. If two devices are grounded via different paths and a communication cable between the devices provides another ground connection between them, a ground loop results. Slightly different potentials in the two power system grounds can cause circulating currents in this ground loop if there is indeed a complete path. Even if there is not a complete path, the insulation that is preventing current flow may flash over because the communication circuit insulation levels are generally quite low.

Likewise, very low magnitudes of circulating current can cause serious noise problems. The best solution to this problem in many cases is to use optical couplers in the communication lines, thereby eliminating the ground loop and providing adequate insulation to withstand transient overvoltages. When this is not practical, the grounded conductors in the signal cable may have to be supplemented with heavier conductors or better shielding. Equipment on *both* ends of the cable should be protected with arresters in addition to the improved grounding because of the coupling that can still occur into signal circuits.

7.3.7 Insufficient neutral conductor

Switch-mode power supplies and fluorescent lighting with electronic ballasts are becoming more and more prevalent in commercial environments (Sec. 5.7). The high third-harmonic content present in these load currents can have a very important impact on the required neutral conductor rating for the supply circuits.

Third-harmonic currents in a balanced system appear in the zero-sequence circuit. This means that third-harmonic currents from three single-phase loads will add in the neutral, rather than cancel as is the case for the 60-Hz current. In typical commercial buildings with a diversity of switched-mode power supply loads, the neutral current is typically in the range 140 to 170 percent of the fundamental frequency phase current magnitude.

The CBEMA has recognized this concern and has prepared a brief to alert the industry to problems caused by harmonics from computer power supplies. The possible solutions to neutral conductor overloading include the following:

- Run a separate neutral conductor for each phase in a three-phase circuit that serves single-phase nonlinear loads.
- When a shared neutral must be used in a three-phase circuit with single-phase nonlinear loads, the neutral conductor capacity should be approximately double the phase conductor capacity.
- Delta-wye transformers designed for nonlinear loads can be used to limit the penetration of high neutral currents. These transformers should be placed as close as possible to the nonlinear loads (e.g., in the computer room). The neutral conductors on the secondary of each separately derived system must be rated based on the expected neutral current magnitudes.
- Filters to control the third-harmonic current that can be placed at the individual loads are becoming available. These will be an alternative in existing installations where changing the wiring may be an expensive proposition.

7.4 Solutions to Wiring and Grounding Problems

7.4.1 Proper grounding practices

Figure 7.3 illustrates the basic elements of a properly grounded electrical system. The important elements of the electrical system grounding are described below.

7.4.2 Ground electrode (rod)

The grounding rod provides the electrical connection from the power system ground to earth. The item of primary interest in evaluating the adequacy of the ground rod is the resistance of this connection. There are three basic components of resistance in a grounding rod:

1. *Electrode resistance.* Resistance due to the physical connection of the grounding wire to the grounding rod.
2. *Rod-earth contact resistance.* Resistance due to the interface between the soil and the rod. This resistance is inverse-

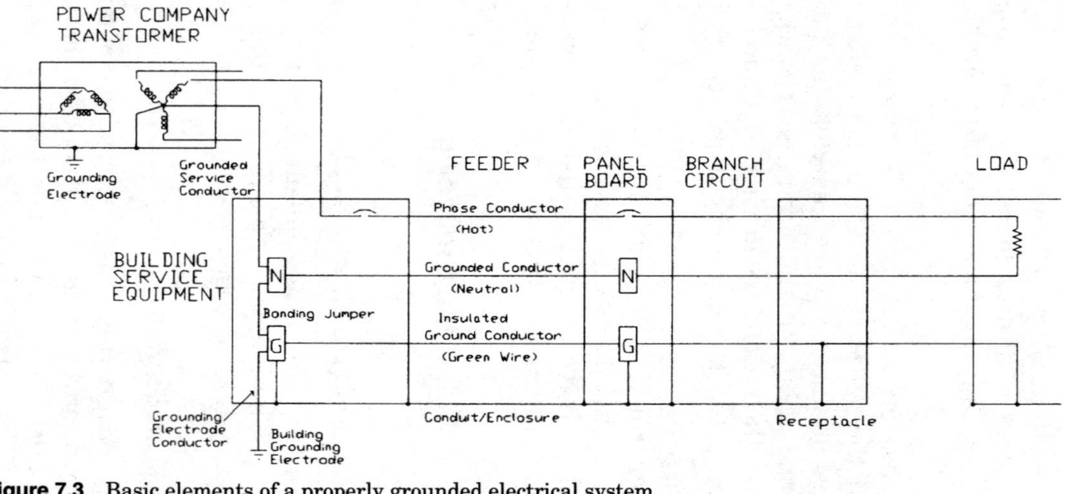

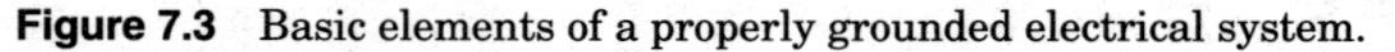

Figure 7.3 Basic elements of a properly grounded electrical system.

ly proportional to the surface area of the grounding rod (i.e., more area of contact means lower resistance).

3. *Ground resistance.* Due to the resistivity of the soil in the vicinity of the grounding rod. The soil resistivity varies over a wide range, depending on the soil type and moisture content.

The resistance of the ground rod connection is important because it influences transient voltage levels during switching events and lightning transients. High-magnitude currents during lightning strokes result in a voltage across the resistance, raising the ground reference for the entire facility. The difference in voltage between the ground reference and true earth ground will appear at grounded equipment within the facility, and this can result in dangerous touch potentials.

7.4.3 Service entrance connections

The primary components of a properly grounded system are found at the service entrance. The neutral point of the supply power system is connected to the grounded conductor (neutral wire) at this point. This is also the one location in the system (except in the case of a separately derived system) where the grounded conductor is connected to the ground conductor (green wire) via the bonding jumper. The ground conductor is also connected to the building grounding electrode via the grounding-electrode conductor at the service entrance. For most effective grounding, the grounding-electrode conductor should be exothermically welded at both ends.

The grounding-electrode conductor is sized based on guidelines in the *National Electrical Code* (Section 250-94). Table 250-94 (reproduced in Table 7.2) from the *NEC* provides the basic guidelines.

There are a number of options for the building grounding electrode. It is important that all of the different grounding electrodes used in a building are connected together at the service entrance. The following are permissible for grounding electrodes:

- *Underground water pipe.* (See *NEC* Table 250-94 for grounding-electrode conductor requirements for connection to the neutral bus.)
- *Building steel.* (See *NEC* Table 250-94 for grounding-electrode conductor requirements for connection to the neutral bus or the underground water pipe.)

TABLE 7.2 Grounding-Electrode Conductor for AC Systems

Size of largest service entrance conductor or equivalent area for parallel conductors		Size of grounding electrode conductor	
Copper	Aluminum or copper-clad aluminum	Copper	Aluminum or copper-clad aluminum
2 or smaller	0 or smaller	8	6
1 or 0	2/0 or 3/0	6	4
2/0 or 3/0	4/0 or 250 MCM	4	2
Over 3/0–350 MCM	Over 250 MCM–500 MCM	2	0
Over 350 MCM–600 MCM	Over 500 MCM–900 MCM	0	3/0
Over 600 MCM–1100 MCM	Over 900 MCM–1750 MCM	2/0	4/0
Over 1100 MCM	Over 1750 MCM	3/0	250 MCM

- *Ground ring.* A ground ring can be used in addition to building steel to provide a better equipotential ground for the grounding electrode. It is connected to the main grounding electrode with a conductor that is not larger than the ground ring conductor.
- *Concrete-encased electrode.* This can serve a similar purpose to a ground ring and is connected to the main grounding electrode with a conductor that has a minimum size #4 AWG.
- *Ground rod.* The ground rod is connected to the main building grounding electrode with a conductor that has a minimum size #6 AWG.

Throughout the system, a safety ground must be maintained to ensure that all exposed conductors that may be touched are kept at an equal potential. This safety ground also provides a ground fault return path to the point where the power source neutral conductor is grounded. The safety ground can consist of the conduit itself or the conduit and a separate conductor (ground conductor or green wire) in the conduit. This safety ground originates at the service entrance and is carried throughout the building.

7.4.4 Panel board

The *panel board* is the point in the system where the various branch circuits are supplied by a feeder from the service entrance. The panel board provides breakers in series with the phase conductors, connects the grounded conductor (neutral) of the branch circuit to that of the feeder circuit, and connects the ground conductor (green wire) to the feeder ground conductor, conduit, and enclosure. It is important to note that there should not be a neutral-to-ground connection at the panel board. This neutral-to-ground connection is prohibited in the *NEC* since it would result in load return currents flowing in the ground path between the panel board and the service entrance. In order to maintain an equipotential grounding system, the ground path should not contain any load return current. Also, fault currents would split between the neutral conductor and the ground return path. Protection is based on the fault current flowing in the ground path.

7.4.5 Isolated ground

The noise performance of the supply to sensitive loads can sometimes be improved by providing an isolated ground to the load. This is done using isolated ground receptacles, which are orange in color. If an isolated ground receptacle is being used downline from the panel board, the isolated ground conductor is not connected to the conduit or enclosure in the panel board, but only to the ground conductor of the supply feeder (Fig. 7.4). The conduit is the safety ground in this case and is connected to the enclosure. A separate conductor can also be used for the safety ground in addition to the conduit. This technique is described in *NEC* Article 274, Exception 4 on receptacles. It is not described as a grounding technique.

The isolated ground receptacle is orange in color for identification purposes. This receptacle does not have the ground conductor connected to the receptacle enclosure or conduit. The isolated ground conductor may pass back through several panel boards without being connected to local ground until grounded at the service entrance or other separately derived ground. The use of isolated ground receptacles requires careful wiring practices to avoid unintentional connections between the isolated ground and the safety ground. In general, dedicated branch circuits accomplish the same objective as isolated ground receptacles without the concern for complicated wiring.

A special case of isolated grounds is used for some hospital equipment grounding. These procedures are described in the *NEC* and in the *White Book* (IEEE Standard 602).

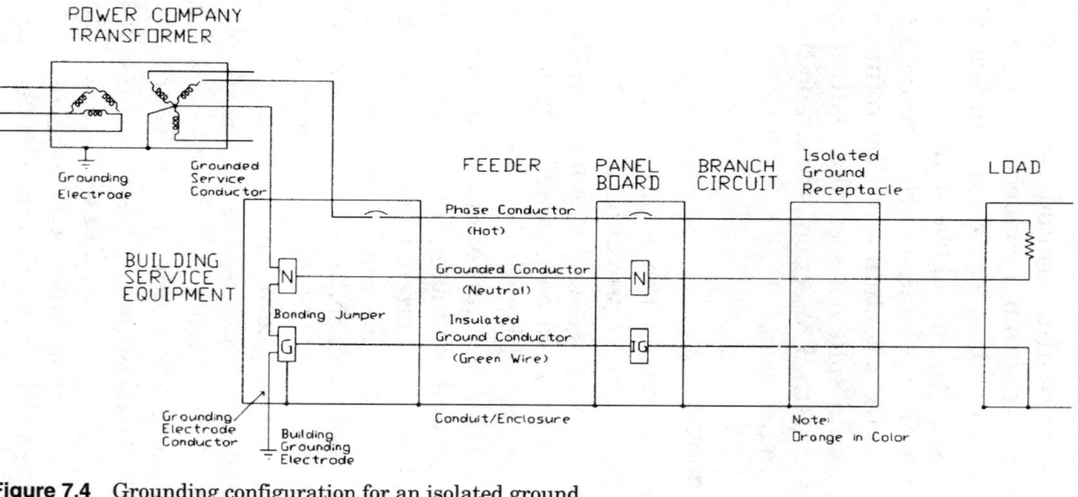

Figure 7.4 Grounding configuration for an isolated ground.

7.4.6 Separately derived systems

A *separately derived system* has a ground reference which is independent from other systems. A common example of this is a delta-wye grounded transformer (Fig. 7.5). The wye-connected secondary neutral is connected to local building ground (not a separate ground rod) to provide a new ground reference independent from the rest of the system. The point in the system where this new ground reference is defined is like a service entrance in that the system neutral is connected to the grounded conductor (neutral wire), which is connected to the ground conductor with a bonding jumper.

Separately derived systems are used to provide a local ground reference for sensitive loads. The local ground reference can have significantly reduced noise levels as compared to the system ground if an isolation transformer is used to supply the separately derived system. An additional benefit is that neutral currents are localized to the load side of the separately derived system. This can help reduce neutral current magnitudes in the overall system when there are large numbers of single-phase nonlinear loads.

7.4.7 Grounding techniques for signal reference

Most of the grounding requirements described above deal with the concerns for safety and proper operation of protective devices. Grounding is also used to provide a signal reference point for equipment exchanging signals over communication or control circuits within a facility. The requirements for a signal reference ground are often significantly different from the requirements for a safety ground. However, the safety ground requirements must always be considered first whenever designing a grounding scheme.

The most important characteristic of a signal reference ground is that it must have a low impedance over a wide range of frequencies. One way to accomplish this (at least for low frequencies) is to use an adequately sized ground conductor. Conduit is particularly bad for a signal reference ground because it relies on continuity of connections and the impedance is high relative to the phase and neutral conductors. Undersized ground conductors have the same problem of high impedance. For reducing power quality problems, the ground conductor should be at least the

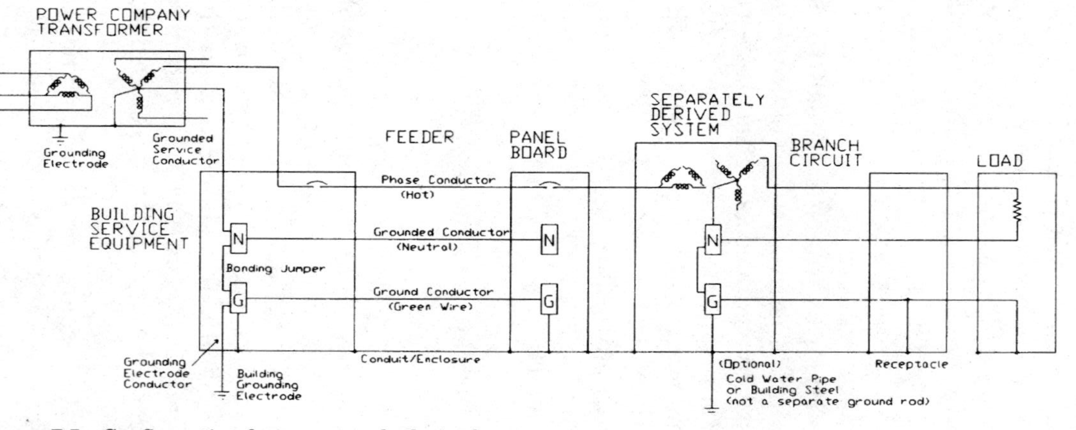

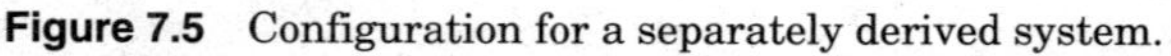

Figure 7.5 Configuration for a separately derived system.

same size as the phase conductors and the neutral conductor (the neutral conductor may need to be larger than the phase conductors in some special cases involving nonlinear single-phase loads).

As frequency increases, the wavelength becomes short enough to cause resonances for relatively short lengths of wire. A good rule of thumb is that when the length of the ground conductor is greater than one-twentieth of the signal wavelength, the ground conductor is no longer effective at that frequency. Since the grounding system is more complicated than a simple conductor, there is actually a complicated impedance vs. frequency characteristic involved (Fig. 7.6).

One way to provide a signal reference ground to sensitive equipment that is effective over a wide range of frequencies (0 to 30 MHz) is to use a signal reference grid or zero reference grid (Fig. 7.7). This technique uses a rectangular mesh of copper wire with about 2-ft spacing. It is commonly applied in large data processing equipment rooms. Even if a portion of the conductor system is in resonance at a particular frequency, there will always be other paths of the grid that are not in resonance due to the multiple paths available for current to flow. When using a signal reference grid, the enclosure of each piece of equipment must still be connected to a single common ground via the ground conductor (*NEC* requirement). The enclosures may also be connected to the closest interconnection of the grid to provide a high-frequency, low-impedance signal reference. Figure 7.6 illustrates the effect of the signal reference grid on the overall ground impedance vs. frequency characteristic.

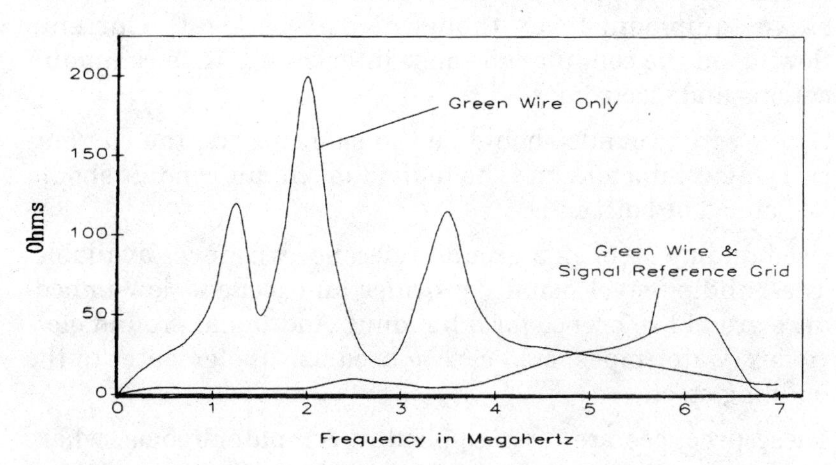

Figure 7.6 Effect of signal reference grid on ground impedance.

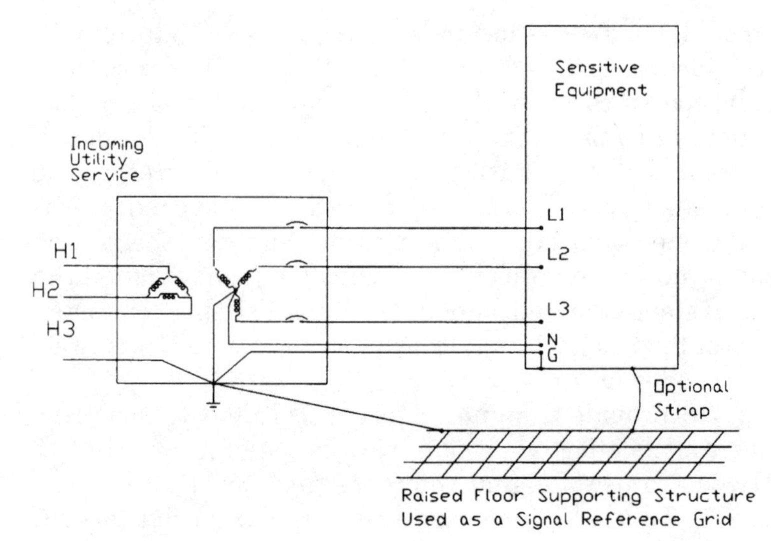

Figure 7.7 Use of a signal reference grid.

7.4.8 More on grounding for sensitive equipment

The following practices are appropriate for any installation with equipment that may be sensitive to noise or disturbances introduced due to coupling in the ground system:

- Whenever possible, use individual branch circuits to power sensitive equipment. Individual branch circuits provide good isolation for high-frequency transients and noise.
- Conduit should never be the sole source of grounding for sensitive equipment (even though it may be legal). Currents flowing on the conduit can cause interference with communications and electronics.
- Green wire grounds should be the same size as the current-carrying conductors and the individual circuit conduit should be bonded at both ends.
- Use building steel as a ground reference whenever available. The building steel usually provides an excellent, low-impedance ground reference for a building. Additional ground electrodes (water pipes, etc.) can be used as supplemental to the building steel.
- These practices are often applied in computer rooms, where the frequency response of the grounding system is even more

important due to communication requirements between different parts of a computer system.

- Either install a signal reference grid under a raised floor or use the raised floor as a signal reference grid. This is not a replacement for the safety ground, but augments the safety ground for noise reduction.
- Addition of a transient suppression plate at or near the power entry point (with the power cabling laid on top of it) to provide a controlled capacitive and magnetic coupling noise bypass between building reinforced steel and the electrical ground conductors.

7.4.9 Summary of wiring and grounding solutions

The grounding system should be designed to accomplish these minimum objectives:

1. There should never be load currents flowing in the grounding system under normal operating conditions. There are likely to be low currents in the grounding system due to the connection of protective devices and coupling between line and ground (in fact, if the ground current is actually zero, there is probably an open ground connection). However, these currents should be negligible compared with load currents.
2. There should be, as near as possible, an equipotential reference for all devices and locations in the system.
3. To avoid excessive touch potential safety risks, all equipment and enclosures should be connected to the equipotential grounding system.

The most important implications resulting from these objectives are:

1. There can only be one neutral-to-ground bond for any subsystem. A separately derived system may be created with a transformer, allowing establishment of a new neutral-to-ground bond.
2. There must be sufficient interconnections in the equipotential plane to achieve a low impedance over a wide frequency range.
3. All equipment and enclosures should be grounded.

Chapter

8

Monitoring Power Quality*

Power quality investigations often require monitoring to identify the exact problem and then to verify the solutions which are implemented. Before embarking on extensive monitoring programs, it is important to develop an understanding of the customer facility, equipment being affected, wiring and grounding practices, and operating considerations. Often, power quality problems can be solved without extensive monitoring by asking the right questions when talking to the customer and performing an initial site survey.

8.1 Site Survey

The initial site survey should be designed to obtain as much information as possible about the customer facility and the problems being experienced. Specific information that should be obtained at this stage includes:

1. Nature of the problems (data loss, nuisance trips, component failures, control system malfunctions, etc.).
2. Characteristics of the sensitive equipment experiencing problems (equipment design information or at least application guide information).
3. When do problems occur?

*This chapter was written by Christopher J. Melhorn.

4. Coincident problems or known operations (e.g., capacitor switching) that occur at the same time.
5. Possible sources of power quality variations within the facility (motor starting, capacitor switching, power electronic equipment operation, arcing equipment, etc.).
6. Existing power conditioning equipment being used.
7. Electrical system data (one-line diagrams, transformer sizes and impedances, load information, capacitor information, cable data, etc.).

Once these basic data have been obtained through discussions with the customer, a site survey should be performed to verify the one-line diagrams, electrical system data, wiring and grounding integrity, load levels, and basic power quality characteristics. Data forms that can be used for this initial verification of the power distribution system are provided in Figs. 8.1 to 8.4.

8.2 Detailed Power Quality Monitoring

Power quality monitoring beyond the initial site survey is performed to characterize power quality variations at specific system locations over a period of time. The monitoring requirements depend on the particular problem that is being experienced. For instance, problems that are caused by voltage sags during remote faults on the utility system could require monitoring for a significant length of time because system faults are probably rare. If the problem involves capacitor switching, it may be possible to characterize the conditions over the period of a couple days. Harmonic distortion problems should be characterized over a period of at least one week to get a picture of how the harmonics vary with load changes. The following sections describe important aspects of the power quality monitoring effort.

8.2.1 Choosing a monitoring location

It is best to start monitoring as close as possible to the sensitive equipment being affected by power quality variations. It is important that the monitor sees the same variations that the sensitive equipment sees. High-frequency transients, in particular, can be significantly different if there is significant separation between the monitor and the affected equipment. Another important location is the main service entrance. Transients and

Supply Transformer Data:

Manufacturer: ______

Connection: ______

kVA Rating: ______

Primary Voltage: ______

Secondary Voltage: ______

Taps: ______

Tap Position: ______

Test Data:

Primary Voltages:		Primary Currents:	
A-B	______	A	______
B-C	______	B	______
C-A	______	C	______
A-N	______	Neutral	______
B-N	______	Ground	______
C-N	______		

Secondary Voltages:		Secondary Currents:	
A-B	______	A	______
B-C	______	B	______
C-A	______	C	______
A-N	______	Neutral	______
B-N	______	Ground	______
C-N	______		

N-G Bond? ______

Figure 8.1 Form for recording supply transformer test data.

voltage variations measured at this location can be experienced by all of the equipment in the facility. This is also the best indication of disturbances caused by the utility system (although it is still very possible that disturbances at the service entrance are caused by events occurring within the facility).

8.2.2 Disturbance recording form

It is important that the customer maintain a log detailing equipment problems that occur during the measurement period with a

Panel Identification: ______________________

Location: ______________________

Voltages:		Feeder Currents:	
A-B	________	A	________
B-C	________	B	________
C-A	________	C	________
A-N	________	Neutral	________
B-N	________	Ground	________
C-N	________		
		Feeder Wire Sizes:	
N-G Bond?	________	Phase	________
		Neutral	________
		Ground	________

Comments: ______________________

Figure 8.2 Form for recording feeder circuit test data (from panel).

disturbance recording form (Fig. 8.5). This will permit correlation of disturbances and system switching events with actual equipment power quality problems. The log should also indicate any major changes in the system configuration that are implemented during the measurement period (power factor correction capacitors, circuit configurations, new equipment, etc.). There will also be many disturbances recorded that do not result in any direct effect on customer equipment. It is important to distinguish these disturbances from the events that actually cause problems.

8.2.3 Disturbance monitor connections

The recommended practice is to provide input power to the monitor from a circuit other than the circuit to be monitored. Some manufacturers include input filters and/or surge suppressors on their power supplies that can alter disturbance data if the monitor is powered from the same circuit that is being monitored.

The grounding of the power disturbance monitor is an important consideration. The disturbance monitor will have a ground

Panel Identification: ____________________

Location: ____________________

Circuit Identifier	Breaker	Phase A	Phase B	Phase C	Neutral	Ground	Loads Served

Figure 8.3 Form for recording branch circuit test data (from panel).

Branch Circuit Identification: ____________________

Location: ____________________

Equipment/ Location	Volts Ph-Ph	Volts Ph-N	Volts N-G	Load Current	Ground Z	Neutral Z

Figure 8.4 Form for recording test data at individual loads.

- Date of Disturbance:
- Time of Disturbance:
- Company:
- Address:
- Contact Name:
- Phone #:
- Brief Description of Disturbance:
- Equipment Category:
- Equipment Type:
- Manufacturers:
- Equipment Limitation:
- Cost of Equipment Failure:
- Cost of Downtime:

Figure 8.5 Sample disturbance recording form.

connection for the signal to be monitored and a ground connection for the power supply of the instrument. Both of these grounds will be connected to the instrument chassis. For safety reasons, both of these ground terminals should be connected to earth ground. However, this has the potential of creating ground loops if different circuits are involved.

Safety comes first. Therefore, both grounds should be connected whenever there is a doubt about what to do. If ground loops can be a significant problem such that transient currents might damage the instruments or invalidate the measurements, it may be possible to power the instrument from the same line that is being monitored (check to make sure there is no signal conditioning in the power supply). Alternatively, it may be possible to connect just one ground (signal to be monitored) and place the instrument on an insulating mat. Appropriate safety practices such as using insulated gloves when operating the instrument must be employed if it is possible for the instrument to rise in potential with respect to other apparatus and ground references with which the operator can come into contact.

8.2.4 Setting monitor thresholds

Disturbance monitors are designed to detect conditions that are abnormal. Therefore, it is necessary to define the range of conditions that can be considered normal. Some disturbance monitors have preselected (default) thresholds that can be used as a starting point.

The best approach for selecting thresholds is to match them with the specifications of the equipment that is affected. This may not always be possible due to a lack of specifications or application guidelines. An alternative approach is to set the thresholds fairly tight for a period of time (collect a lot of disturbance data) and then use the data collected to select appropriate thresholds for longer duration monitoring.

8.2.5 Quantities to measure

When monitoring power disturbances, it is usually sufficient to monitor system voltages. This is not adequate for harmonic measurements. In order to characterize harmonic concerns, it is critical to measure both voltages and currents. If you have to choose one or the other, the currents are generally more important.

Current measurements are used to characterize the generation of harmonics by nonlinear loads on the system. Current measurements at individual loads are valuable for determining these harmonic generation characteristics. Current measurements on feeder circuits or at the service entrance characterize a group of loads or the entire facility as a source of harmonics. Current measurements on the distribution system can be used to characterize groups of customers or an entire feeder.

Voltage measurements help characterize the system response to the generated harmonic currents. Resonance conditions will be indicated by high harmonic voltage distortion at specific frequencies. In order to determine system frequency response characteristics from measurements, voltages and currents must be measured simultaneously. In order to measure harmonic power flows, all three phases must be sampled simultaneously.

8.2.6 Interpreting the measurement results

In order to analyze power quality problems using measurements, it is important to be able to correlate the characteristics of a disturbance with possible causes of the disturbance. This requires a knowledge of the characteristics which are typical for different types of disturbances. The waveforms and information presented in this book are designed to provide the background needed to interpret a variety of different power quality variations. Once the cause of a disturbance is understood, the impacts on equipment and possible solutions must be determined. There is not always a direct cause-and-effect relationship between the disturbance and the effect on equipment (long-term degradation, interaction with controls can influence problems). This can make evaluation of the impacts and development of appropriate solutions more difficult.

8.2.7 Finding the source of a disturbance

The first step in identifying the source of a disturbance is to correlate the disturbance waveform with possible causes, as outlined above. Once a category for the cause has been determined (e.g., load switching, capacitor switching, remote fault condition, recloser operation, etc.), the identification becomes more straightforward. The following general guidelines can be used:

- High-frequency voltage variations will be limited to locations close to the source of the disturbance. Low-voltage (600 V and below) wiring often damps out high-frequency components very quickly due to circuit resistance so that these frequency components will only appear when the monitor is located close to the source of the disturbance.
- Power interruptions close to the monitoring location will cause a very abrupt change in the voltage. Power interruptions remote from the monitoring location will result in a decaying voltage due to stored energy in rotating equipment and capacitors.
- The highest harmonic voltage distortion levels will occur close to capacitors that are causing resonance problems. In these cases, a single frequency usually dominates the voltage harmonic spectrum.

8.3 Power Quality Measurement Equipment

Power quality problems encompass a wide range of disturbances and conditions on the system. They include everything from very fast transient overvoltages (microsecond time frame) to long duration outages (hours or days time frame). Power quality problems also include steady-state phenomena such as harmonic distortion, and intermittent phenomena, such as voltage flicker. Definitions for the different categories were presented in Chap. 2. This wide variety of conditions which make up "power quality" makes the development of standard measurement procedures and equipment very difficult.

8.3.1 Types of instruments

Although instruments have been developed which measure a wide variety of disturbances, a number of different instruments are generally necessary, depending on the phenomena being investigated. Basic categories of instruments which may be applicable include:

- Wiring and grounding test devices
- Multimeters
- Oscilloscopes
- Disturbance analyzers
- Harmonic analyzers/spectrum analyzers
- Combination disturbance and harmonic analyzers
- Flicker meters
- Energy monitors

The following sections discuss the application and limitations of these different instruments. Besides these instruments, which measure steady-state signals or disturbances on the power system directly, other instruments can be used to help solve power quality problems by measuring ambient conditions:

- Infrared meters can be very valuable in detecting loose connections and overheating conductors. An annual procedure of checking the system in this manner can help prevent power quality problems due to arcing, bad connections, and overloaded conductors.
- Noise problems related to electromagnetic radiation may require measurement of field strengths in the vicinity of affected equipment. Magnetic gauss meters are used to measure magnetic field strengths for inductive coupling concerns. Electric field meters can measure the strength of electric fields for electrostatic coupling concerns.

- Static electricity meters are special purpose devices to measure static electricity in the vicinity of sensitive equipment. Electrostatic discharge (ESD) can be an important cause of power quality problems in some types of electronic equipment.

Regardless of the type of instrumentation needed for a particular test, a number of important factors should be considered when selecting the instrument. Some of the more important factors include:

- Number of channels (voltage and/or current)
- Temperature specifications of the instrument
- Ruggedness of the instrument
- Input voltage range (e.g., 0 to 600 V)
- Power requirements
- Ability to measure three-phase voltages
- Input isolation (isolation between input channels and from each input to ground)
- Ability to measure currents
- Housing of the instrument (portability, rack-mount, etc.)
- Ease of use (user interface, graphics capability, etc.)
- Documentation
- Communication capability (modem, network interface)
- Analysis software

The flexibility (comprehensiveness) of the instrument is also important. The more functions that can be performed with a single instrument, the fewer instruments will be required. Recognizing that there is some crossover between the different instrument categories, the basic categories of instruments for direct measurement of power signals are discussed on the following pages.

8.3.2 Wiring and grounding testers

The great majority of power quality problems reported by customers are caused by problems with wiring and/or grounding within the facility. These problems can be identified by visual inspection of wiring, connections, and panel boxes, and also with special test devices for detecting wiring and grounding problems. Important capabilities for a wiring and grounding test device include

- Detection of isolated ground shorts and neutral-ground bonds
- Ground impedance and neutral impedance measurement or indication
- Detection of open grounds, open neutral, or open hot wire
- Detection of hot/neutral reversals or neutral/ground reversals

Three-phase wiring testers should also test for phase rotation and phase-to-phase voltages. These test devices can be simple and provide an excellent initial test for circuit integrity. Many problems can be detected without the requirement for detailed monitoring using expensive instrumentation.

8.3.3 Multimeters

After initial tests of wiring integrity, it may also be necessary to make quick checks of the voltage and/or current levels within a facility. Overloading of circuits, under- and overvoltage problems, and unbalances between circuits can be detected in this manner. These measurements just require a simple multimeter. Signals to check include

- Phase-to-ground voltages
- Phase-to-neutral voltages
- Neutral-to-ground voltages
- Phase-to-phase voltages (three-phase system)
- Phase currents
- Neutral currents

The most important factor to consider when selecting and using a multimeter is the method of calculation used in the meter. All of the commonly used meters are calibrated to give an rms indication for the measured signal. However, a number of different methods are used to calculate the rms value. The three most common methods are:

1. *Peak method.* Assuming the signal to be a sinusoid, the

meter reads the peak of the signal and divides the result by 1.414 (square root of 2) to obtain the rms.

2. *Averaging method.* The meter determines the average value of a rectified signal. For a clean sinusoidal signal (signal containing only one frequency), this average value is related to the rms value by a constant.
3. *True rms.* The root mean square (rms) value of a signal is a measure of the heating which will result if the voltage is impressed across a resistive load. One method of detecting the true rms value is to actually use a thermal detector to measure a heating value. More modern digital meters use a digital calculation of the rms value by squaring the signal on a sample-by-sample basis, averaging over the period, and then taking the square root of the result.

These different methods all give the same result for a clean, sinusoidal signal but can give significantly different answers for distorted signals. This is very important because significant distortion levels are common, especially for the phase and neutral currents within the facility. Table 8.1 can be used to better illustrate this point.

Each waveform in Table 8.1 has an rms value of 1.0 pu (100.0 percent). The corresponding measured values for each type of meter are displayed under the associated waveforms, normalized to the true rms value.

8.3.4 Oscilloscopes

An oscilloscope is valuable when performing real-time tests. Looking at the voltage and current waveforms can provide much information about what is happening, even without performing detailed harmonic analysis on the waveforms. You can get the magnitudes of the voltages and currents, look for obvious distortion, and detect any major variations in the signals. There are numerous makes and models of oscilloscopes to choose from. A digital oscilloscope with data storage is valuable because the waveform can be saved and analyzed. Oscilloscopes in this category often have waveform analysis capability (energy calculation, spectrum analysis) also. In addition, digital oscilloscopes can usually be obtained with communications so that waveform data can be uploaded to a PC for additional analysis with a software package.

TABLE 8.1 Comparison of Meter Reading for Various Waveforms

		Meter Type		
		True RMS	Peak Method	Average Responding
		Circuit Type		
		RMS Converter	Peak / 1.414	Sine Avg. X 1.11
Sine Wave		100 %	100 %	100 %
Square Wave		100 %	82 %	110 %
Triangle Wave		100 %	121 %	96 %
ASD Current		100 %	127 %	86 %
PC Current		100 %	184 %	60 %
Light Dimmer		100 %	113 %	84 %

The latest developments in oscilloscopes are hand-held instruments with the capability to display waveforms as well as performing some signal processing. These are quite useful for power quality investigations because they are very portable and can be operated like a volt-ohm meter, but yield much more information. These are ideal for initial plant surveys. A typical device is shown in Fig. 8.6. This particular instrument also has the capability to analyze harmonics and permits connection with PCs for further data analysis and inclusion into reports as illustrated.

8.3.5 Disturbance analyzers

Disturbance analyzers and disturbance monitors form a category of instruments which have been developed specifically for power quality measurements. They typically measure a wide variety of system disturbances from very short duration transient voltages to long-duration outages or undervoltages. Thresholds can be set and the instruments left unattended to record disturbances over a period of time. The information is most commonly

Figure 8.6 A hand-held power quality monitoring instrument. (*Photograph courtesy of Fluke Corp.*)

recorded on a paper tape but many devices have attachments so that it can be recorded on disk as well.

The devices basically fall into two categories:

1. *Conventional analyzers* that summarize events with specific information such as over/undervoltage magnitudes, sags/surge magnitude and duration, transient magnitude and duration, etc.
2. *Graphics-based analyzers* that save and print the actual waveform along with the descriptive information which would be generated by one of the conventional analyzers.

It is often difficult to determine the characteristics of a disturbance or a transient from the summary information available from conventional disturbance analyzers. For instance, an oscillatory transient cannot be effectively described by a peak and a duration. Therefore, it is almost imperative to have the waveform capture capability of a graphics-based disturbance analyzer

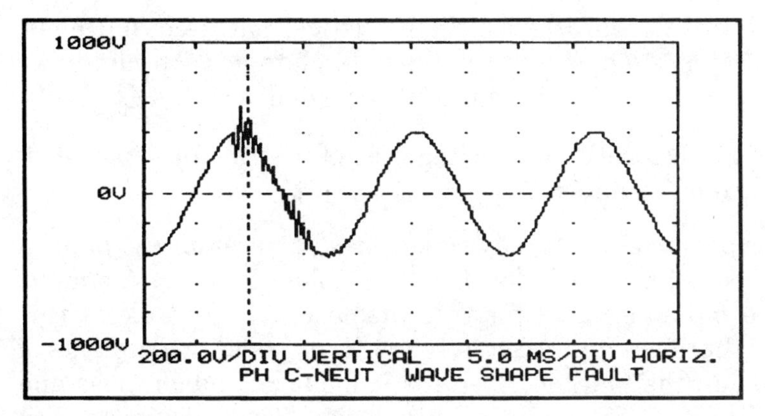

Figure 8.7 Graphics-based analyzer output.

for detailed analysis of a power quality problem (Fig. 8.7). However, a simple conventional disturbance monitor can be valuable for initial checks at a problem location.

8.3.6 Spectrum analyzers and harmonic analyzers

Instruments in the disturbance analyzer category have very limited harmonic analysis capabilities. Some of the more powerful analyzers have add-on modules that can be used for computing fast Fourier transform (FFT) calculations to determine the lower-order harmonics. However, any significant harmonic measurement requirements will demand an instrument that is designed for spectral analysis or harmonic analysis. Important capabilities for useful harmonic measurements include

- Capability to measure both voltage and current simultaneously so that harmonic power flow information can be obtained.
- Capability to measure both magnitude and phase angle of individual harmonic components (also needed for power flow calculations).
- Synchronization and a sampling rate fast enough to obtain accurate measurement of harmonic components up to at least the 37th harmonic (this requirement is a combination of a high sampling rate and a sampling interval based on the 60-Hz fundamental).

- Capability to characterize the statistical nature of harmonic distortion levels (harmonics levels change with changing load conditions and changing system conditions).

There are basically three categories of instruments to consider for harmonic analysis:

1. *Simple meters.* It may sometimes be necessary to make a quick check of harmonic levels at a problem location. A simple, portable meter is ideal for this purpose. At the time of this writing, there are approximately four hand-held instruments of this type on the market. Each instrument has advantages and disadvantages in its operation and design. These devices generally use microprocessor-based circuitry to perform the necessary calculations to determine individual harmonics up to the 50th harmonic, as well as the rms, the THD, and the telephone influence factor (TIF). Some of these devices can calculate harmonic powers (magnitudes and angles) and can upload stored waveforms and calculated data to a PC.
2. *General-purpose spectrum analyzers.* Instruments in this category are designed to perform spectrum analysis on waveforms for a wide variety of applications. They are general signal analysis instruments. The advantage of these instruments is that they offer very powerful capabilities for a reasonable price since they are designed for a broader market than simply power system applications. The disadvantage is that they are not designed specifically for sampling 60-Hz waveforms and, therefore, must be used carefully to assure accurate harmonic analysis. There are a wide variety of instruments in this category.
3. *Special-purpose power system harmonic analyzers.* Besides the general-purpose spectrum analyzers described above, there are also a number of instruments and devices designed specifically for power system harmonic analysis (Fig. 8.8). These are based on the FFT with sampling rates specifically designed for determining harmonic components in power signals. They can generally be left in the field and include communications capability for remote monitoring.

8.3.7 Combination disturbance and harmonic analyzers

The most recent instruments combine limited harmonic-sampling and energy-monitoring functions with complete distur-

+1.0KA

0A

-1.0KA

AMPS

500.0A/DIV VERTICAL 3.3MS/DIV HORIZ.

PHASE A CURRENT SPECTRUM 11:49:06 AM
Fundamental amps: 103.8 A rms
Fundamental freq: 60.0 Hz

HARM	PCT	SINE PHASE	HARM	PCT	SINE PHASE
FUND	100.0%	10°	2nd	1.1%	78°
3rd	3.9%	-122°	4th	0.5%	167°
5th	82.8%	-125°	6th	1.7%	-56°
7th	77.5%	79°	8th	1.2%	131°
9th	7.6%	-80°	10th	0.7%	112°
11th	46.3%	-52°	12th	1.0%	-48°
13th	41.2%	149°	14th		
15th	5.7%	-26°	16th	0.3%	172°
17th	14.2%	19°	18th	0.4%	78°
19th	9.7%	-145°	20th	0.4%	-138°
21st	2.3%	19°	22nd	0.5%	-14°
23rd	1.5%	-148°	24th	0.5%	89°
25th	2.5%	108°	26th	0.7%	-135°
27th	0.9%	-29°	28th	0.3%	9°
29th	2.0%	-29°	30th	0.2%	55°
31st	2.0%	169°	32nd	0.3%	149°
33rd	0.5%	-19°	34th	0.4%	-61°
35th	0.3%	-147°	36th	0.1%	25°
37th	0.8%	75°	38th	0.3%	148°
39th	0.5%	-58°	40th		
41st	0.6%	-100°	42nd		
43rd	0.7%	114°	44th	0.1%	113°
45th	0.4%	-59°	46th	0.1%	-32°
47th	0.2%	165°	48th		
49th	0.4%	44°	50th	0.3%	144°
ODD	130.9%		EVEN	3.0%	
THD:	130.9%				

Figure 8.8 Harmonic analyzer output.

bance-monitoring functions as well. The output is graphically based and the data are remotely gathered over telephone lines into a central database. Statistical analysis can then be performed on the data. The data are also available for input and manipulation into other programs such as spreadsheets and other graphical output processors.

One example of such an instrument is shown in Fig. 8.9. This instrument is designed for both utility and end-user applica-

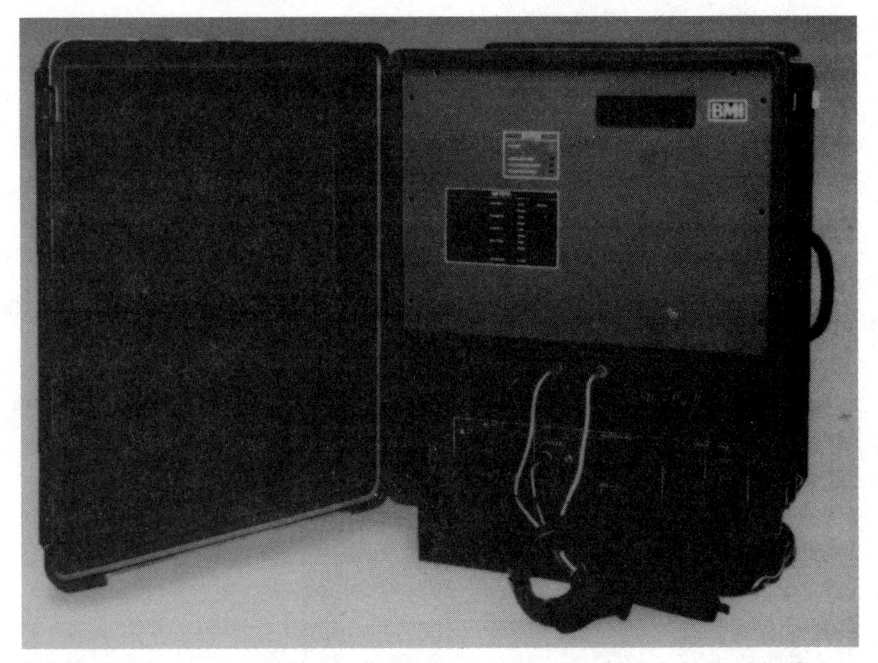

Figure 8.9 A power quality monitoring instrument capable of monitoring disturbances, harmonics, and other steady-state phenomena on both utility systems and end-user systems. (*Photograph courtesy of Basic Measuring Instruments.*)

tions, being mounted in a suitable enclosure for installation outdoors on utility poles. It monitors three-phase voltages and currents (plus neutrals) simultaneously, which is very important for diagnosing power quality problems. The instrument captures the raw data and saves the data in internal storage for remote downloading. Off-line analysis is performed with powerful software that can produce a variety of outputs such as that shown in Fig. 8.10. The top chart shows a typical result for a voltage sag. Both the rms variation for the first 0.8 s and the actual waveform for the first 175 ms are shown. The middle chart shows a typical wave fault capture from a capacitor-switching operation. The bottom chart demonstrates the capability to report harmonics of a distorted waveform. Both the actual waveform and the harmonic spectrum can be obtained. Another device is shown in Fig. 8.11. This is a load analysis monitoring system that also has the capability to capture disturbances and analyze harmonics. It is made for permanent installations in industrial facilities and can be distributed throughout the plant on motor control centers and subpanels.

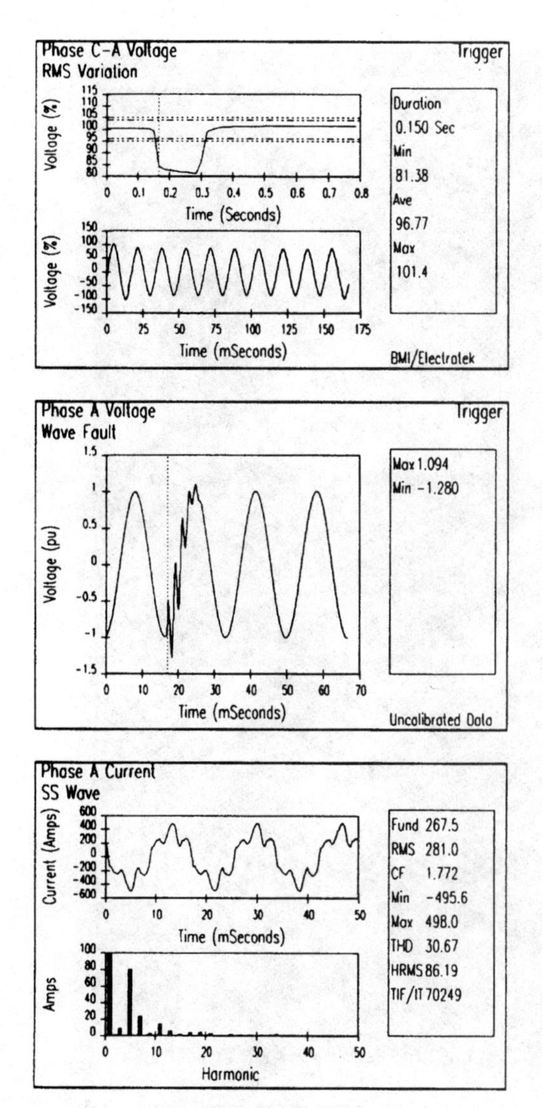

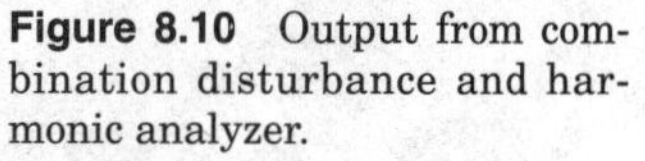
Figure 8.10 Output from combination disturbance and harmonic analyzer.

Thus, while only a few short years ago, power quality monitoring was a rare feature to be found in instruments, it is becoming much more commonplace in commercially available equipment.

8.3.8 Flicker meters

Voltage flicker is a term for small but rapid changes in the supply voltage. This term is used because of the effect of these variations (typically in the range 1 to 30 Hz) on electric lamps

Figure 8.11 A combination load and power quality monitor for permanent installation in industrial facilities. (*Photograph courtesy of Square D Company.*)

perceived as flickering by the human eye. Arc furnaces are one major cause of voltage flicker, but other loads such as rock crushers and tire testers with randomly varying characteristics can also cause similar problems.

When measuring flicker, an instrument must measure the rms value of the disturbance voltage, or the envelope of the 60-Hz voltage. It is obtained by demodulating the envelope from the 60-Hz carrier. It is also important to calculate the dominant frequency in the flicker signal since different frequencies are perceived differently by the human eye. A number of utilities have built their own flicker meters based on analog circuitry, and commercial flicker meters are now available. Some are self-contained meters while others are based on PCs.

8.3.9 Transducer requirements

Monitoring of power quality on power systems often requires transducers to obtain acceptable voltage and current signal levels. Voltage monitoring on secondary systems can usually be performed with direct connections but even these locations require current transformers (CTs) for the current signal. Many power quality monitoring instruments are designed for input voltages up to 600 V rms and current inputs up to 5 A rms. Voltage and current transducers must be selected to provide these signal levels. Two important concerns must be addressed in selecting transducers:

1. *Signal levels.* Signal levels should use the full scale of the instrument without distorting or clipping the desired signal.
2. *Frequency response.* This is particularly important for transient and harmonic distortion monitoring, where high-frequency signals are particularly important.

These concerns, and transducer installation considerations, are discussed below.

8.3.10 Signal levels

Careful consideration to sizing of voltage transducers (VTs) and CTs is required to take advantage of the full resolution of the instrument without clipping the measured signal. Improper sizing can result in damage to the transducer or monitoring instrument.

Digital monitoring instruments incorporate the use of analog-to-digital (A/D) converters. These A/D boards convert the analog signal received by the instrument from the transducers into a digital signal for processing. To obtain the most accurate representation of the signal being monitored, it is important to use as much of the full range of the A/D board as possible. The noise level of a typical A/D board is approximately 33 percent of the full-scale bit value (5 bits for a 16-bit A/D). Therefore, as a general rule, the signal that is input to the instrument should never be less than one-eighth of the full scale value, so that it is well above the noise level of the A/D board. This can be accomplished by selecting the proper transducers.

Voltage transducers. VTs should be sized to prevent measured disturbances from inducing saturation in the VT. For transients, this generally requires that the knee point of the transducer saturation curve be at least 200 percent of nominal system voltage.

Example. When monitoring on a 12.47-kV distribution feeder and measuring line to ground, the nominal voltage across the primary of the voltage transducer is 7200 V rms.

A VT ratio of 60:1 will produce an output voltage on the VT of 120 V rms (170 V peak) for 7200 V rms input. Therefore, if the full range value of the instrument is 600 V rms and the instrument incorporates a 16-bit A/D, 13+ bits of the A/D will be used.

It is always good practice to incorporate some allowance in the calculations for overvoltage conditions. The steady-state voltage should not be right at the full scale value of the monitoring instrument. If an overvoltage occurred, the signal would be clipped by the A/D board, and the measurement would be useless. Allowing for a 150 percent overvoltage is suggested. This can be accomplished by changing the input scale on the instrument, or sizing the VT accordingly.

Current transformers. Selecting the proper transducer for currents is more difficult. The current in any system changes more often and with greater magnitude than the voltage. Most power quality instrument manufacturers supply CTs with their equipment. These CTs come in a wide range of sizes to accommodate different load levels. The CTs are usually rated for maximum continuous load current.

The proper CT current rating and turns ratio depends on the measurement objective. If fault or inrush currents are of concern, the CT must be sized in the range of 20 to 30 times normal load current. This will result in low resolution of the load currents and inability to accurately characterize load current harmonics.

If harmonics and load characterization are important, CTs should be selected to accurately characterize load currents. This permits evaluation of load response to system voltage variations and accurate calculation of load current harmonics.

Example. The desired current signal to the monitoring instrument is 1 to 2 A rms. Assuming a 1-A value, the optimum CT ratio for an average feeder current of 120 A rms is 120:1. Manufacturer's data commonly list CT turns ratios on a base of 5 A rather than 1 A. The primary rating for a given CT is calculated as follows:

$$\mathrm{CT}_{\mathrm{PRI}} = \frac{I_{\mathrm{PRI}}\mathrm{CT}_{\mathrm{SEC}}}{I_{\mathrm{SEC}}} = \frac{120 \cdot 5}{1} = 600 \tag{8.1}$$

Thus, a 600:5 CT should be specified.

8.3.11 Frequency response

Transducer frequency response characteristics can be illustrated by plotting the ratio correction factor (RCF), which is the ratio of the expected output signal (input scaled by turns ratio) to the actual output signal, as a function of frequency.

Voltage transducers. The frequency response of a standard metering class VT depends on the type and burden. In general, the burden should be a very high impedance[1] (Figs. 8.12 and 8.13). This is generally not a problem with most monitoring equipment available today. Power quality monitoring instruments, digital multimeters, oscilloscopes, and other instruments all present a very high impedance to the transducer. With a high impedance burden, the response is usually adequate to at least 5 kHz.

Some substations use capacitively coupled voltage transformers (CCVTs) for voltage transducers. These *should not* be used for general power quality monitoring. There is a low-voltage transformer in parallel with the lower capacitor in the capaci-

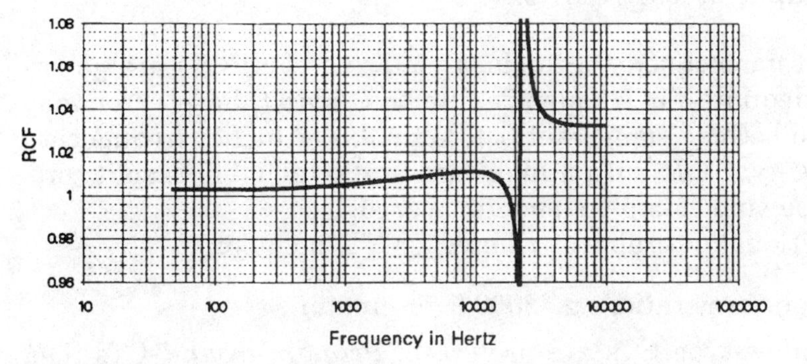

Figure 8.12 Frequency response of a standard VT with 1-MΩ burden.

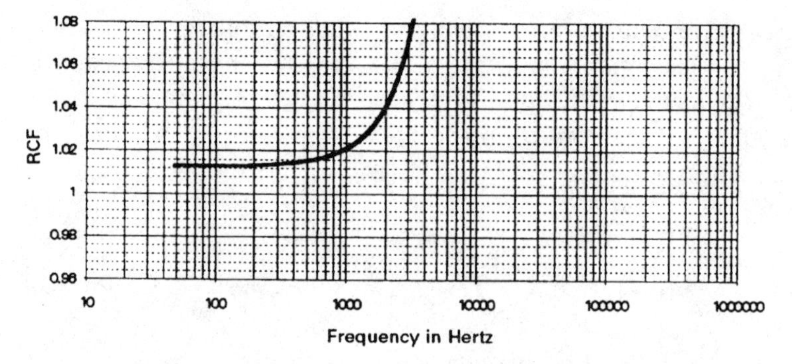

Figure 8.13 Frequency response of a standard VT with 100-Ω burden.

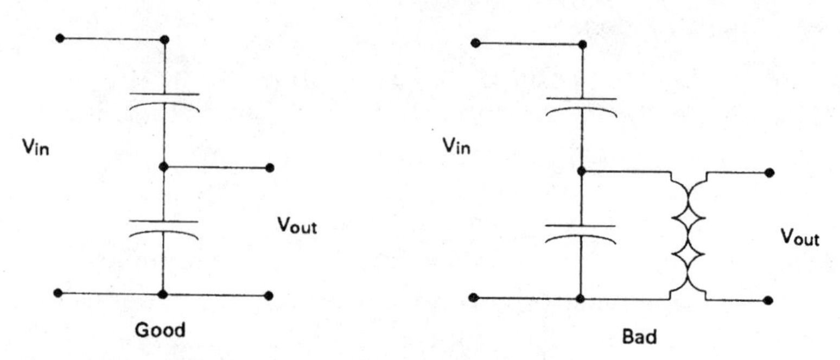

Figure 8.14 Capacitively coupled voltage dividers.

tive divider. This configuration results in a circuit that is tuned to 60 Hz and will not provide accurate representation of any higher-frequency components.

Measuring very high frequency components in the voltage requires a capacitive divider or pure resistive divider. Figure 8.14 illustrates the difference between a CCVT and a capacitive divider. Special-purpose capacitor dividers can be obtained for measurements requiring accurate characterization of transients up to at least 1 MHz.

Current transformers. Standard metering class CTs are generally adequate for frequencies up to 2 kHz (phase error may start to become significant before this). For higher frequencies, *window-type CTs* with a high turns ratio (doughnut, split core, bar type, and clamp-on) should be used.[2]

Additional desirable attributes for CTs include:

1. Large turns ratio, e.g., 2000:5, or greater.
2. Window-type CTs are preferred. *Primary-wound* CTs (i.e.,

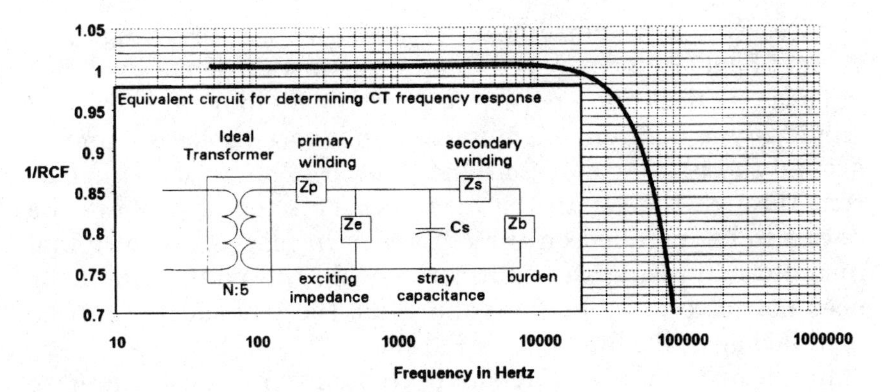

Figure 8.15 Frequency response of a window-type CT.

CTs in which system current flows through a winding) may be used, provided that the number of turns is less than five.

3. Small remnant flux, e.g., ≤ 10 percent of the core saturation value.
4. Large core area. The more steel used in the core, the better the frequency response of the CT.
5. Secondary winding resistance and leakage impedance as small as possible. As shown in Fig. 8.15, this allows more of the output signal to flow into the burden, rather than the stray capacitance and core-exciting impedance.

8.3.12 Installation considerations

Monitoring on the distribution primary requires both voltage and current transducers. Selection of the best combination of these transducers depends on a number of factors.

- Monitoring location (substation, overhead, underground, etc.)
- Space limitations
- Ability to interrupt circuit for transducer installation
- Need for current monitoring

Substation transducers. Usually, existing substation CTs and VTs (except CCVTs) can be used for power quality monitoring.

Utility overhead line locations. For power quality monitoring on distribution primary circuits, it is often desirable to use a transducer that could be installed without taking the circuit out of service. Recently, transducers for monitoring both voltage and current have been developed that can be installed on a live line.

These devices incorporate a resistive divider-type VT and window-type CT in a single unit. A split-core choke is clamped around the phase conductor, and is used to shunt the line current through the CT in the insulator. This method allows the device to be installed on the cross-arm in place of the original insulator. By using the split-core choke, the phase conductor does not have to be broken, and thus, the transducers can be installed on a live line.

Initial tests indicated adequate frequency response for these transducers. However, field experience with these units has shown that the frequency response, even at 60 Hz, is dependent on current magnitude, temperature, and secondary cable length. This makes this type of device difficult to use for accurate power quality monitoring. Care must be exercised in matching these transducers to the instruments.

In general, all primary sites should be monitored with metering class VTs and CTs to obtain accurate results over the required frequency spectrum. Installation requires a circuit outage but convenient designs can be developed for pole top installations to minimize the outage.

Another option for monitoring primary sites involves monitoring at the secondary of an unloaded distribution transformer. This gives accurate results up to at least 3 kHz. This option does not help with the current transducers, but it is possible to get by without the currents at some circuit locations (e.g., end of the feeder). This option may be particularly attractive for underground circuits where the monitor can be installed on the secondary of a pad-mounted transformer.

Primary-wound CTs are available from a variety of CT manufacturers. Reference 2 concludes that any primary-wound CT with a single turn, or very few turns, should have a frequency response up to 10 kHz.

End-user (secondary) sites. Transducer requirements at secondary sites are much simpler. Direct connection for the voltage is possible for 120/208 V rms or 277/480 V rms systems. This permits full utilization of the instrument's frequency response capability.

Currents can be monitored either with metering CTs (e.g., at the service entrance) or with clamp-on CTs (at locations within the facility). Clamp-on CTs are available in a wide range of turns ratios. The frequency range is usually published by the manufacturer.

8.3.13 Summary of transducer recommendations

Table 8.2 describes different monitoring locations and the different types of transducers that are adequate for monitoring at these locations.

Table 8.3 describes the different power quality phenomena and the proper transducers to measure that type of power quality problem. Tables 8.2 and 8.3 should be used in conjunction

TABLE 8.2 VT and CT Options

Location	VT	CT
Substation	Metering VTs Special-purpose capacitive or resistive dividers Calibrated bushing taps	Metering CTs Relaying CTs
Overhead lines	Metering VTs	Metering CTs
Underground locations	Metering VTs Pad-mounted transformer Special-purpose dividers	Metering CTs
Secondary sites		
Service entrance	Direct connection	Metering CTs Clamp-on CTs
In facility	Direct connection	Clamp-on CTs

TABLE 8.3 VT and CT Requirements

Concern	VTs*	CTs
Voltage variations	Standard metering	Standard metering
Harmonic levels	Standard metering	Window-type
Low-frequency transients (switching)	Standard metering with high-knee-point saturation	Window-type
High-frequency transients (lighting)	Capacitive or resistive dividers	Window-type

*VTs are usually not required at locations below 600 V rms nominal.

	Wiring Problems	Impulses & Transients	Voltage Variations	Interruptions	Harmonics	Flicker	Noise	Electrostatic Discharge
Wiring and Grounding Testers	■							
Multimeters	■		■					
Oscilloscopes		■						
Disturbance Analyzers		■	■	■			■	
Harmonic Analyzers					■			
Flicker Meters						■		
Infrared Detectors	■							
Gauss Meters							■	
Field Strength Meters							■	
Static Meters								■

Figure 8.16 Power quality measurement equipment capabilities.

with each other to determine the best transducer for a given application.

8.4 Summary of Equipment Capabilities

Figure 8.16 summarizes the capabilities of the previously described metering instruments as they relate to the various categories of power quality variations.

8.5 References

1. D. A. Douglas, "Potential Transformer Accuracy at 60 Hz Voltages above and below Rating and at Frequencies above 60 Hz," presented at the IEEE Power Engineering Society Summer Meeting, Minneapolis, July 13–18, 1980.
2. D. A. Douglas, "Current Transformer Accuracy with Asymmetric and High Frequency Fault Currents," *IEEE Transactions on Power Apparatus on Systems,* vol. PAS-100, no. 3, March 1981.

8.6 Bibliography

C. J. Cokkinides, L. E. Banta, A. P. Meliopoulos, "Transducer Performances for Power System Harmonic Measurements," in *Proceedings of the International Conference on Harmonics,* Worcester, Mass., October 1984.

"Computation of Current Transformer Transient Performance," *IEEE Transactions on Power Delivery,* vol. PWRD-3, no. 4, October 1988.

A. N. Greenwood, *Electrical Transients in Power Systems,* 2d ed., John Wiley & Sons, New York, 1991, Chap. 18.

E. L. McShane and M. E. Colbaugh, *Advance Current and Voltage Transformers for Power Distribution Systems,* EPRI Report EL-6289, 1989.

Index

ABOUT THE AUTHORS

ROGER C. DUGAN is senior consultant with Electrotek Concepts, Inc. He has over 20 years of experience in the power industry, with a focus on computer simulations of power systems. He is an expert in the application of distribution switchgear for overcurrent protection, surge arresters, power factor correction capacitors, step-voltage regulators, and transformers.

MARK F. MCGRANAGHAN is the general manager of power systems engineering at Electrotek Concepts, Inc. He has worked with electric utilities and end users nationwide in analyzing power quality problems and devising solutions. He teaches seminars on a wide range of topics related to power quality, and is very active in the development of national and international power quality standards.

H. WAYNE BEATY is editor or managing editor of numerous power industry publications, including *Electric Light & Power* magazine. He is also coeditor of *Standard Handbook for Electrical Engineers*, Thirteenth Edition, which is available from McGraw-Hill.